GENETICS? NO PROBLEM!

GENETICS? NO PROBLEM!

Kevin O'Dell

School of Life Sciences, University of Glasgow
Glasgow, UK

Library of Congress Cataloging-in-Publication data applied for:

ISBN (Hardback): 9781118833889
ISBN (Paperback): 9781118833872

A catalogue record for this book is available from the British Library.

Wiley also publishes its books in a variety of electronic formats. Some content that appears in print may not be available in electronic books.

Cover image: Mouse © Frances Jones
        Psychofish © Hayley Patterson
        Zombiology Institute © Time-Tastical Productions
        Doug and Kevin © Time-Tastical Productions
        Hazardous cupboard © Kevin O'Dell
        Pig © Kevin O'Dell
        Petri Dish © Kevin O'Dell
        Tiger © Del Bonds
        Strawberry Biomass Graph © Kevin O'Dell
        All other images from Public Domain
Cover design by Dan Jubb

Set in 10.5/13pt Palatino by SPi Global, Pondicherry, India
Printed and bound in Singapore by Markono Print Media Pte Ltd

10 9 8 7 6 5 4 3 2

Contents

Section 3: Advanced

Foreword

I start teaching my first year genetics course at University College London by saying 'I am a geneticist, and my job is to make sex boring'. The students look rather baffled, but after twenty or so lectures, I can tell that they heartily agree. It's not that genetics is of its nature uninteresting – for as every biologist knows there are plenty of good stories about inherited disease, mutation, inbreeding, evolution and much more – but the unfortunate fact is that genetics has been, even from its earliest days, a quantitative subject. For too many of today's students, the Q word makes them quail, and as I am fighting through the analysis of pedigrees, linkage mapping, population genetics, heritability, inbreeding coefficients and the like, I can see at least part of my audience's eyes glaze over.

This book should brighten them up again. Instead of just slogging through reams of figures in a lecture or a tutorial, each of its many exercises is embedded in a narrative, from illegitimacy to murder and from Bengal Tigers to guinea pigs. *Genetics? No Problem!* is not about the tedious accumulation of fact (although any science without fact is a house built on sand), but about a series of worked problems that lay the foundations upon which a sturdy factual edifice can be built. The book goes from the elementary to the advanced, and from plant flower colours to the mythic inhabitants of Titan – one of Saturn's moons – taking in some equally fanciful Scottish creatures (albino haggis anyone?) on the way.

Some of the examples may stretch a reader's imagination – but that is what modern biology itself has done, again and again. This book will work best in consort with examples from the real rather than the imagined world and many of its exercises are close to those used by investigators of viral infection, circadian rhythms and much more. As a catalyst for discussions and exercises about what was described to me when I was a student as 'genetical thinking' (although it was never explained as clearly as this book does quite what that might be), even a few small aliquots of its contents will go a long way.

Professor Steve Jones FRS, University College London, UK

Preface

The analysis and interpretation of data is fundamental to understanding genetics and forms a compulsory part of the undergraduate genetics curriculum. Among the key skills that a genetics student requires are (a) an ability to design and understand experimental strategies and (b) problem-solving skills to interpret those experimental results and data. In my experience, students often find this rather difficult and a little bit scary. Problems in standard textbooks are usually apologetically tacked onto the end of chapters and are always presented in a rather dry way. So, to help engage the students, I have embedded the problems in short narratives.

This is exclusively a problem-solving genetics textbook. There is no formal descriptive material included, other than an introduction to provide clear instructions on how to use the book. I want to enthuse, challenge and entertain the reader in an informative manner, and a key to this is getting the reader to think for themselves. The genetics problems are embedded in stories that will, I hope, be amusing and entertaining.

The book is divided into three sections: introductory, intermediate and advanced. For first level students there will be short genetics problems embedded in a wide range of scenarios, such as murder mysteries and, as the book progresses, the stories will get longer and the science will get progressively more complex so that final year students will be identifying genetic disease in obscure organisms as well as designing and testing treatments and cures.

The problem-solving approach is in the format of real scientific research, as many questions are about the students analysing some data, proposing a model that interprets that data, and then devising an experimental strategy that tests their model. I hope that the book will encourage students to take responsibility for their own learning by requiring them to go and find appropriate information for themselves.

Kevin O'Dell

Acknowledgements

Nobody in science works in glorious isolation, and this book could never have been written without the support of family, friends and colleagues. When I arrived in Glasgow over twenty-five years ago, I was fortunate to have the chance to work with two very talented researchers, Howy Jacobs and Kim Kaiser, and two hugely inspirational lecturers, Roger Sutcliffe and Richard Wilson. I owe the four of you a great deal. The Genetics Degree team at Glasgow have been a pleasure to work with over the years, and their contribution to this book cannot be underestimated. In particular, comments from Mark Bailey, Joe Gray and Darren Monckton have always been welcome, even though it may not have felt like this at the time. In addition, the support and encouragement from Rob Aitken in the School of Life Sciences at Glasgow is hugely appreciated. I should also thank the students. It's been great fun and without you there would be many more typos and incomprehensible questions. Finally, and most importantly, I'd like to thank Christine, Adam and Annie, who make everything so worthwhile.

Kevin O'Dell

How to Use this Book

When I was at school I was lucky enough to have a couple of biology teachers who could make even the driest bit of biology absolutely fascinating. Somehow they could dress the material in a context that would bring it spectacularly to life. I can't exactly explain why an essay entitled 'Why can't amoeba grow to the size of a number 207 double-decker bus?' is much more interesting than the more traditional 'Why are amoeba small?', especially as 95% of the answer is going to be the same, but I would work a lot harder at the former than the latter. Mr Quilley and Ms Grounds have a lot to answer for!

The other thing that I've discovered after 25 years of teaching genetics at the University of Glasgow, is that the very best students are the ones who can work things out for themselves. So over the years we've been progressively moving away from traditional lectures and trying different ways of getting students to do things for themselves. One of the best ways of doing that is problem solving, and that is what this book is all about.

How do the very best students become the next generation of world-class researchers? It's not because they can remember some complex formula, or the precise spelling of the Latin name for an obscure disease, or even because they can draw the exact chemical structure of every amino acid. It's because they have a good broad understanding of their subject, the ability to find information from appropriate sources, to have ideas and hypotheses about how things might work, and the energy, intelligence and enthusiasm to test those ideas. In the end, it's all about problem solving.

This book is all about developing your problem-solving skills. For those of you on a standard three-year undergraduate degree programme, you'll find ten problems targeted at level one, ten more for level two and another ten pitched at final year students. Many of these are based on problem-solving sessions we've had with the genetics students at the University of Glasgow, whilst others are slightly modified versions of their exams (though don't let that put you off).

The key is to have a go at the problems and see how far you get. Don't be disappointed if you don't understand something and don't be afraid of getting things wrong. Nobody understands everything first time, and if any teaching isn't challenging, it's probably not actually worth doing! It may well be the case that you'll have to look things up, but that's exactly what the career scientists would do. If you know more at the end of a question than you did at the start, then progress has been made.

Sometimes there is a very specific answer. Given a precise set of data, for example, two genes are 7.2 map units apart (although an answer of 7.18 may well be equally as correct), but sometimes it isn't quite as straightforward as that. Just like in the real research world, in this book when you have a set of data, you'll often be asked to come up with some ideas of what's going on. You will often be asked to 'speculate', and as long as your interpretation fits the data, then you're on the right track. Of course there could be multiple possible interpretations of the data, so the key to good research is then coming up with several ways of investigating whether your model is correct. Again there may be several plausible answers, each of which is kind of correct, but some aspects of an answer may be better than others. I've provided some ideas of what I think is going on but see if you can come up with some more original speculation, and then devise a way of testing it!

So, enjoy the book. I have.

Section 1

Introductory

Chapter 1.1

Grandma's Secret

It's a cold and wet summer day. So at your mother's request you're clearing most of your stuff out of the loft before you go to university to study genetics. It's not a very exciting job but it is about to get a whole lot more interesting.

In the corner of the loft you find the old suitcase shown in Fig. 1.1.1 and, as the padlock is broken, you carefully open it. Folded neatly inside you find a collection of newspapers and some old family photographs dating from the 1960s. You guess they must have belonged to your grandma who would have been a teenager back then. All the great events of the 1960s are recorded, including Sidney Poitier winning an Oscar (April 1964), the death of racing driver Jim Clark (April 1968) and the first moon landing (July 1969). There are also a lot of teenage 'popular music' magazines with stories about the great bands of that era such as the Beatles, the Rolling Stones and The Fridge.

Figure 1.1.1   Grandma's suitcase.

Genetics? No Problem!, First Edition. Kevin O'Dell.
© 2017 John Wiley & Sons Ltd. Published 2017 by John Wiley & Sons Ltd.

At the bottom of the suitcase you find a large old envelope. You open it and discover some newspaper cuttings about Paul Cool, lead singer and occasional drummer of The Fridge, who is shown in Fig. 1.1.2. He was the bad boy of the British 1960s popular music scene, and most of the articles are about him driving cars into swimming pools, being rude to old ladies and not wearing a tie on a Sunday. But one article catches your eye. In a cutting dated 30th July 1966, there is the headline 'The Fridge's Cool denies being the father of little Brenda'. What surprises you is that in the accompanying picture, the teenage girl holding baby Brenda looks remarkably like a young version of your Grandma. It also occurs to you that your mum's older sister is called Brenda, and she was born in the spring of 1966.

You've always secretly admired Aunt Brenda. She is pretty cool in a bohemian kind of way. However, your parents aren't quite so positive. 'Embarrassing for a women of her age,' is the best your mum can say. 'Trashy,' is your father's favourite phrase, though you've always wondered whether he secretly fancies her. Could Aunt Brenda actually be the daughter of Paul Cool?

Figure 1.1.2   Paul Cool. 1960s rock icon, lead singer and occasional drummer of The Fridge. This photograph was also found in Grandma's suitcase.

You continue searching and find some medical records inside a second envelope with what seems to be blood group data. This information is reproduced in Table 1.1.1. The blood group data is accompanied by a letter stating 'Paternity unresolved' and is signed by Doctor Robert.

Table 1.1.1 Blood groups of a mother (your Grandma), a child (Aunt Brenda) and an alleged father (Paul Cool). This data was found in the second envelope.

Subject	Blood group
Mother (your Grandma)	AB
Child (Aunt Brenda)	A
Alleged father (Paul Cool)	B

You look at the data, and at first something doesn't seem quite right. However, you've studied the ABO blood group system at school and so you check your notes.

Question 1: What is the genetic basis of the ABO blood group system? (10%)

Question 2: Briefly discuss the limitations of testing relatedness using ABO blood groups. (10%)

Question 3: What could the genotypes of your Grandma, Paul Cool and Aunt Brenda be, if your Grandma and Paul Cool are indeed Aunt Brenda's parents? (10%)

Question 4: What could the genotypes of your Grandma, Paul Cool and Aunt Brenda be, if your Grandma is Aunt Brenda's mother but Paul Cool is not Aunt Brenda's father? (10%)

Question 5: Why is Doctor Robert's interpretation correct? (5%)

By now you're desperate to discover the truth, and you soon realise that DNA testing is the only solution. Over the next few weeks you secretly and covertly collect DNA samples from every member of your family. Hairbrushes disappear on a regular basis, and your mother is particularly suspicious when you pour her a large glass of Chateau Lafite 1999 and suggest she takes the weight off her feet while you do the washing up, by hand. After all, it is only ten o'clock on a Saturday morning.

Collecting Paul Cool's DNA proves more problematic. Much to your family's surprise you volunteer to help at *Codgers*, the local old peoples home. You soon discover which room Paul Cool lives in and pay the old rock icon a visit. In a moment of complete madness and no little bravery you take a swab from his bedpan and escape.

Being a student you don't have much money, so when you send your collection of DNA samples to Alec's Genetic Testing Company you request they only sample a single short tandem repeat (STR) rather than embark upon the full STR testing process.

Question 6: What are short tandem repeats (STRs)? (10%)

Question 7: Briefly describe an experiment that would enable you to determine the length of an STR. (10%)

Question 8: Why are STRs used to determine relatedness, especially paternity? (10%)

Three weeks later an unmarked brown envelope arrives. You open it with great excitement. Inside is a table of data revealing the lengths of a chromosome 13 STR from each member of your family, and from Paul Cool. This data is shown in Table 1.1.2.

Table 1.1.2   The lengths of a chromosome 13 STR for members of your family, and for Paul Cool.

Family member	STR lengths
Grandma	8 & 17
Granddad	11 & 19
Paul Cool	14 & 16
Aunt Brenda	8 & 11
Mum	8 & 19
Dad	12 & 15
You	11 & 15
Your brother	12 & 19
Your sister	15 & 19

Question 9: Who is Aunt Brenda's genetic father? (10%)

Question 10: Draw a family tree to show the apparent relationship between everyone in Table 1.1.2. (10%)

You're so engrossed in the information in the envelope that you don't realise your mother has entered the room and is looking over your shoulder. She asks you what you're doing.

Question 11: What are you going to tell your mother? (5%)

Chapter 1.2

Tiger! Tiger!

You've always dreamed of visiting your cousin in Kolkata and you can't quite believe you're actually on a flight to visit her. More to the point, you can't quite believe your parents have paid for the trip! This is a very special summer in so many respects. You've just finished school and, with any luck and the right exam results, you'll be off to university in September to study genetics.

When you arrive at Netaji Subhash Chandra Bose International Airport your cousin is there to meet you. But she is not alone. 'This is my boss, Professor Bagh,' she says. 'She's wondering whether you'd like to work with us over the summer.' You are totally lost for words. Professor Bagh is a bit of a hero of yours. She runs the Bengal Tiger Conservation Project and nobody knows more about tigers than she does.

The professor smiles at you and says, 'Your cousin tells me you're a budding geneticist. Every good conservation project needs a talented geneticist.' You begin to fear your cousin has been rather exaggerating. You are somewhat reassured when Professor Bagh smiles and says, 'We have several talented geneticists working with us. Would you like to become part of our genetics team?'

Over the next few days, whilst you get used to your new surroundings, you find out as much as you can about the Bengal tiger. You soon discover that even though it is the most numerous of the world's tiger subspecies, there are still fewer than 2500 Bengal tigers in the wild. Its current and historic range is shown in Fig. 1.2.1. Apparently most Bengal tigers live in eastern India, with smaller populations in Bangladesh, Bhutan and Nepal. The Bengal tiger is classified as an endangered species by the International Union for Conservation of Nature (IUCN), and Professor Bagh is developing a strategy to help the species survive.

Genetics? No Problem!, First Edition. Kevin O'Dell.
© 2017 John Wiley & Sons Ltd. Published 2017 by John Wiley & Sons Ltd.

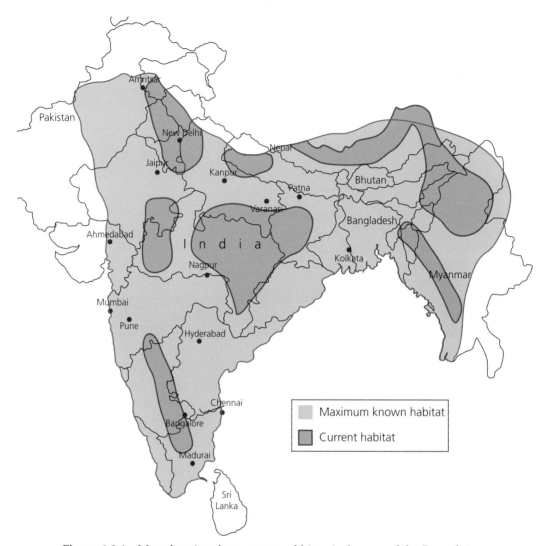

Figure 1.2.1 Map showing the current and historical range of the Bengal tiger.

Your cousin explains that there are several reasons why tiger populations are declining. Apparently, humans cause most of the problems, either through the destruction of habitat or by poaching. This means that the remaining tiger populations are fairly isolated and inbred. So there is a real risk of extinction.

Question 1: Why is inbreeding often a problem for wild animals? (10%)

Question 2: What kind of plan could conservation project leaders, such as Professor Bagh, follow to avoid issues with inbreeding? (10%)

On your first day at work your cousin introduces you to the Professor's team and you're very excited when she asks you to put on some protective clothing. You presume this must be because you're going to start some analysis of tiger DNA samples, so the protective clothing will prevent exposure to the dangerous chemicals you'll be using to do this.

You are therefore slightly surprised when your cousin also hands you a bucket and spade and takes you outside. 'The tigers have been moved to another enclosure,' she says very confidently. 'All you have to do is collect tiger poo and put it in the bucket, whilst I mark the position of the poo on my map so we know which tiger it came from'. She notices the horrified look on your face. 'It's a lot safer than asking a tiger to give you a blood sample,' she says. Whilst you are still somewhat horrified, you have to concede that she has a valid point. Over the summer you become quite an expert at collecting, bagging and labelling tiger poo.

Back in the laboratory, Professor Bagh explains what she wants you to do. 'We occasionally see white Bengal tigers,' she says, 'and we'd like you to help us determine the genetic basis of this variant.' The two versions of Bengal tiger, the common orange form and the very rare white form, are shown in Fig. 1.2.2. Your first thought is that the white tiger is an albino.

Figure 1.2.2   The wild-type orange and rare white varieties of Bengal tiger. Professor Bagh's team use these cartoons to identify individual orange tigers as each individual has a slightly different orange and black pattern. The white tigers only differ in their black patterning.

You ask Professor Bagh to provide you with some tiger pedigree data so you can determine whether the white colour is genetically inherited. One of the tiger pedigrees is shown in Fig. 1.2.3.

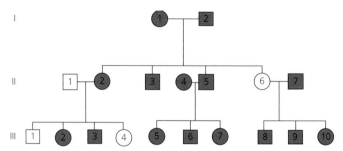

Figure 1.2.3   Inheritance of white and orange Bengal tiger colour in a typical tiger family. By convention, females are circles and males are squares. Generations are indicated by roman numerals (I, II, III and IV), whilst individuals within that generation are numbered 1, 2, etc.

Question 3: Looking at the image of the mutant white tiger in Fig. 1.2.2, give two reasons why the white Bengal tiger is not a true albino. (10%)

Question 4: What is the evidence that the white colour is caused by a recessive mutation? (10%)

Question 5: Redraw the pedigree showing the genotypes of each individual. (10%)

Question 6: There may be some individuals whose genotypes cannot be absolutely determined. Who are they and why can't their genotypes be determined? (10%)

Question 7: How could you determine the precise genotypes of the problem animals identified in your answer to question 6? (10%)

A few days later your cousin rushes into the laboratory in a bit of a panic. 'Your theory must be wrong,' she says. 'Two white Bengal tigers have just given birth to three orange cubs!' You ask her to explain further. 'We've been following tiger movements within the conservation project area for months,' she replies, 'and we're fairly sure Sunetra mated with Arindam.' You note that she uses the phrase *fairly sure* rather than *absolutely certain*, and you realise that it's time to test the parentage of the three orange cubs using DNA profiling. Images of these and of other tigers you have encountered are shown in Fig. 1.2.4.

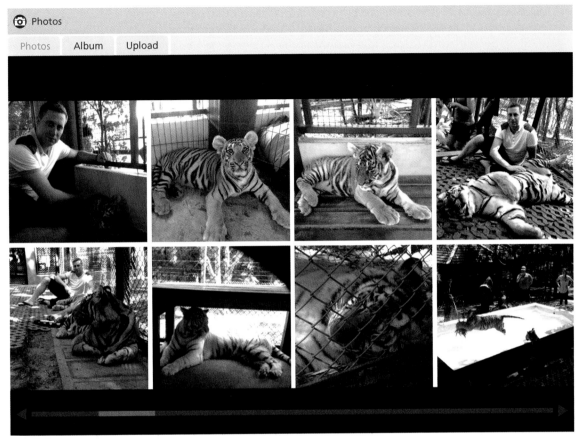

Figure 1.2.4 Images of your time working at the Bengal Tiger Conservation Project with your new best friend Del.

You remember your biology teacher explaining how DNA tests using short tandem repeats (STRs) can be used to determine parentage. So you look at your notes and investigate how STRs are inherited.

Question 8: What is the genetic basis of STRs and how can they be used to determine parentage? (10%)

It's now that you appreciate the value of the work you've been doing in the summer. Your database immediately tells you which samples of tiger poo came from the mother and her three cubs, as well as which samples came from each of the four possible fathers. You collect the appropriate tiger poo samples from the freezer and you prepare DNA samples. For each of the eight tigers, you determine the length of a particularly variable STR on chromosome 4. This data is presented in Table 1.2.1.

Table 1.2.1   The STR status and coat colour of the mother and her three cubs, as well as the STR status and coat colour of the four males who could be the father of the cubs. All of the adults were originally caught in the wild at least a year ago, so their place of capture is also shown. As the gestation period of tigers is approximately 100 days, the cubs must have been conceived since Sunetra arrived at the Bengal Conservation Project.

Name	Colour	Status	Place of capture	STRs
Sunetra	white	mother	West Bengal	22 & 27
Clementine	orange	cub		27 & 31
Satsuma	orange	cub		22 & 31
Tangerine	orange	cub		27 & 34
Arindam	white	father?	West Bengal	22 & 28
Shantanu	orange	father?	West Bengal	20 & 27
Sourav	white	father?	Assam	31 & 34
Utpal	orange	father?	Bihar	18 & 26

Question 9: Who are the parents of the three orange cubs, Clementine, Satsuma and Tangerine? (10%)

Question 10: What is the most plausible explanation for Clementine, Satsuma and Tangerine being orange? (10%)

Chapter 1.3

Anticipation

One of your closest friends is studying medicine and whilst you usually think of them as one of the cleverest (and perhaps most stunningly attractive) people you've ever met, you are surprised that they're struggling with their latest assignment, the diagnostics of Huntington's disease. As you have just started studying genetics, they ask for your help, and to say you are excited is something of an understatement! This is your chance to impress. Your friend gives you a copy of their assignment and you begin studying it with great anticipation.

The assignment concerns a family headed by parents Maureen (aged 35) and Freddie (aged 40). They have three children, Agnes (aged 11), Brian (aged 10) and Caroline (aged 7). However, Freddie has recently been feeling unwell and after tests at the local medical genetics diagnostics unit (see Fig. 1.3.1) he has discovered that he has Huntington's disease. He is naturally worried about his children's future health, particularly as he suspects Agnes and Brian are showing early signs of the disease.

Figure 1.3.1 Your local medical genetics diagnostic unit.

Genetics? No Problem!, First Edition. Kevin O'Dell.
© 2017 John Wiley & Sons Ltd. Published 2017 by John Wiley & Sons Ltd.

A further complication is that Maureen is 16 weeks pregnant, with twins. Fortunately the medics tell him that the twins, a girl they propose to call Doris and a boy they intend to call Edward, seem to be developing well. Maureen and Freddie decide to test their entire family, including the unborn twins, for Huntington's disease.

Your friend shows you the details of their assignment. The data is presented as fragments of DNA from the Huntington gene being separated on an electrophoretic gel and this is reproduced here in Fig. 1.3.2.

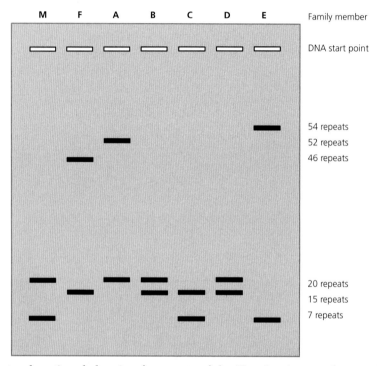

Figure 1.3.2   Electrophoretic gel showing fragments of the Huntingtin gene from each member of the family. The lanes are labelled with the initial letter of the given name of each family member: Maureen (M), Freddie (F), Agnes (A), Brian (B), Caroline (C), Doris (D) and Edward (E).

As you want to impress your friend as much as possible, you decide to include as much detail as you can when you interpret the data.

Question 1: How is Huntington's disease inherited? (10%)

Question 2: What is the molecular basis (DNA change) of the mutation that causes Huntington's disease? (10%)

Question 3: Briefly describe the phenotype of Huntington's disease. (10%)

Question 4: How would you obtain the DNA from each family member? (10%)

Question 5: Briefly describe, using an appropriate figure if necessary, the process by which the DNA test, as depicted in the image in Fig. 1.3.2, is undertaken. (20%)

Question 6: Which members of the family have Huntington's disease? What is the evidence for this? (30%)

Question 7: How reliable is this diagnostic test for Huntington's disease? (10%)

Chapter 1.4

Budgie Hell

You've been working at Mrs Singh's Pet Emporium to earn some money whilst waiting to start studying for your genetics degree. Suddenly you receive a tweet. Mrs Singh is in a flap. It's been a disastrous night. Someone forgot to securely close the doors to the aviaries that house Mrs Singh's prize-winning budgies. When she arrived at her shop early in the morning, a scene of complete and utter chaos confronted her. Budgies were everywhere! The budgies are Mrs Singh's pride and joy and she is mortified. And she blames you!

As an act of contrition, you help her clear up the mess. Mrs Singh has an award-winning collection of budgies and she insists you return each bird to its own aviary. She explains precisely what you need to do in great detail. You realise that she must still be very angry with you. Assigning each budgie to the correct cage isn't the most difficult task you've undertaken as there are simply four colour types: green, blue, yellow and white (see Fig. 1.4.1). You spend the whole day clearing up the mess and even Mrs Singh is forced to concede that you've done a fantastic job. Within a few days Mrs Singh has calmed down and you are delighted when she withdraws her threat to dismiss you.

Figure 1.4.1   The four different colour varieties of budgies.

A couple of months later you're working at the Pet Emporium when Mrs Singh lets out a spectacular scream. She is as angry as you can ever remember. She marches you to the white aviary. It's not immediately obvious what the problem is, but when you peer into a nest you notice a very white budgie tending to her very blue chicks. You choose to bluff your way out of the situation by impressing Mrs Singh with your extensive knowledge of genetics. 'It must be a mutation,' you say very positively. 'Professor Melopsittacus told us all about that in her seminal

Genetics? No Problem!, First Edition. Kevin O'Dell.
© 2017 John Wiley & Sons Ltd. Published 2017 by John Wiley & Sons Ltd.

lecture last week'. Mrs Singh looks on sceptically, but you feel you may just have convinced her. However, as the day progresses and you look at the colours of the chicks in the nests of the other aviaries, you begin to wonder whether Mrs Singh will be suspicious of quite how extraordinarily high the mutation rate would have to be!

Question 1: What is the most plausible reason for the white budgie female having a clutch of seven chicks all of which are blue? (10%)

Over the next few months you make copious notes regarding the parentage of the budgies. You correctly discover that the original four varieties, green, blue, yellow and white, were true-breeding and highly inbred strains. So blue pairs would always give rise to blue, yellow would give rise to yellow, and so on.

To rescue the genetic purity of her four budgie varieties, Mrs Singh removed the offending blue chicks (five females and two males) from the aviary. You kindly offered to look after them, though in fact you gave them to your Aunt Polly as a 40th birthday present. However, you are rather shocked when you visit Aunt Polly six months later and notice that all five blue females are sitting on nests! Surely the budgies realised they were brothers and sisters?!

Question 2: From a genetic perspective, why isn't it usually a good idea for siblings to have offspring? (10%)

When the chicks hatch and fledge you return to Aunt Polly's to look at the young chicks. There are 32 of them, 25 blue and 7 white.

Question 3: What is the most plausible genetic explanation for the colours of the chicks in Aunt Polly's aviary? (10%)

You correctly decide to test your theory statistically with a chi-squared test. The chi-squared formula in Fig. 1.4.2 and Table 1.4.1 should help you. For one degree of freedom, the 5% significance limit is 3.84.

Question 4: Using a chi-squared test, determine whether your explanation of the inheritance of blue and white colours in budgies is likely to be correct. (10%)

$$\chi^2 = \sum \frac{(O-E)^2}{E}$$

Figure 1.4.2   The chi-squared formula.

In fact the blue chicks in the white hen's nest weren't your only problem. You also rescued a clutch of yellow chicks from another white hen's nest. This time you gave them to Aunt Robyn. Once again the yellow brothers and sisters produced offspring. They produced 38 yellow chicks and 10 white chicks.

Table 1.4.1   Use this table to calculate the chi-squared value for the theory you proposed in the answer to question 3.

Colour phenotype	Observed number	Expected number	(O – E)	(O – E)²	$\frac{(O-E)^2}{E}$
Blue	25				
White	7				
Total	32				

Question 5: What is the most plausible genetic explanation for the colours of the chicks in Aunt Robyn's aviary? (5%)

Question 6: Using a chi-squared test, determine whether your explanation of the inheritance of yellow and white colours in budgies is likely to be correct. (5%)

You feel you're really beginning to understand the genetic basis of body colour in budgies. So it is with some confidence that you phone Aunt Phoebe. You gave her twenty-five green chicks that were offspring from four separate nests tended to by white hens in the white aviary. You're pleased there are only four colour types as you're beginning to run out of aunts. The aunts are also beginning to suspect that you're an international budgie smuggler.

At first you're delighted when Aunt Phoebe tells you she has 400 second generation chicks! You know that the more data you have to work with, the more robust any data will be. However, you are shocked when Aunt Phoebe tells you there are 224 green chicks, 83 blue chicks, 70 yellow chicks, and 23 white chicks (see Fig. 1.4.3).

Figure 1.4.3   Budgies living in Aunt Phoebe's aviary.

Question 7: What is the most plausible genetic explanation for the colours of the chicks in Aunt Phoebe's aviary? (10%)

Question 8: Using a chi-squared test, determine whether your explanation of the inheritance of colour in Aunt Phoebe's budgies is likely to be correct. Again there will be 1 degree of freedom and the 5% cut-off for significance is 4.85. (10%)

Question 9: Speculate on the type of proteins the gene(s) might encode. (15%)

Whilst you are at Aunt Phoebe's house trying to explain how she's ended up with so many differently coloured budgies, there is a knock at the door. It's Aunt Phoebe's very old neighbour Archie Optrix with his very strange looking budgie called Harlequin, a sketch of whom is shown in Fig. 1.4.4. You immediately realise that Harlequin is a male budgie as the cere or nostril just above the beak is blue, unlike in females where it is light brown or white, but how did he get that combination of colours?

Figure 1.4.4   A sketch of Harlequin: the strangely pigmented bird that belongs to the neighbour of Aunt Phoebe.

Question 10: Propose a genetic explanation for the colour pattern of Harlequin. (15%)

Friends Reunited

You cannot believe your luck. You only entered the competition to stop your mother from nagging you to do something useful over the summer. She doesn't seem to appreciate that this is your last summer of freedom before you start university. So working hard this summer isn't high on your list of priorities. However, winning the competition changes everything. A month at the Bountiful Lake International Summer School with 11 other prize-winners is a dream come true. Especially as each day a different world-class scientist will visit and tell you all about their amazing research.

You arrive at the summer school in good time and set about meeting some of the other prizewinners. The first international students you meet, Celia and Nicky, are both from England. Apparently they arrived a day early and you are impressed by how much time and effort they've put into exploring the Bountiful Lake resort. They've also already acquired the course details and you make the very sensible decision to join them at the champagne bar on the beach (see Fig. 1.5.1).

Figure 1.5.1   Your first glass of champagne.

Genetics? No Problem!, First Edition. Kevin O'Dell.
© 2017 John Wiley & Sons Ltd. Published 2017 by John Wiley & Sons Ltd.

It's slightly surreal sipping champagne whilst discussing the evolution of mitochondria and you wonder whether university will be like that when you start your genetics degree at the end of the summer. Over the next two hours your new best friends help you learn all about human mitochondrial genetics. Of course you already knew mitochondria were inherited down the female line, but the course notes tell you each human cell has about 500 copies of the 16569 bp circular mitochondrial DNA molecule. The longer you stay in the champagne bar, the more complex mitochondria seem to be.

The weather is amazing, the company is fantastic and you are beginning to think absolutely nothing could top this when, disaster, you step on a jellyfish (see Fig. 1.5.2). You feel a sharp stinging sensation and soon your memory is a complete blank.

Figure 1.5.2   A jellyfish on the beach at Bountiful Lake.

Sometime later you wake up in your hotel room, and the first faces you see belong to the somewhat concerned Celia and Nicky. 'We had no idea you're allergic to jellyfish stings,' they say excitedly and, before you have a chance to tell them it was an even bigger surprise to you, they add, 'You'll need to stay in bed tomorrow and recover, but we'll come and see you as soon as the class is over'. With that they disappear leaving you all alone. Suddenly the summer school doesn't seem quite so exciting after all.

The following evening, Celia and Nicky come to your room as planned. They are gushingly positive about their day and you are insanely jealous. Apparently the day started with all 11 students generating their own DNA sample from their cheek cells. These 11 DNA samples were then placed in a hugely expensive automated Next-Generation DNA Sequencing Machine and just 3 hours later, mitochondrial DNA sequences from the 11 students were available for analysis.

The class was instructed on how to use the mitochondrial DNA sequences from the 11 students to work out how all they were related to each other down their maternal lineage. Of course, most of the 16569 bases were identical, so the class had been advised to focus on 10 polymorphic bases from the non-coding D-loop region of their mitochondrial DNA genomes. The sequences of the 10 polymorphic mitochondrial DNA bases in the 11 students of the international summer school are shown in Table 1.5.1.

Question 1: Define the term 'polymorphism'. (10%)

Question 2: For studies of this type, what are the advantages of looking at non-coding, rather than coding, sequences? (15%)

Table 1.5.1 Mitochondrial DNA sequences of the 11 students in your class at 10 polymorphic sites. Apart from Celia and Nicky, your other classmates are Ahmed from Turkey, Farah from Iraq, Indira from India, Josh from the USA, Li and Zhang from China, Lucia and Sofia from Brazil, and Pablo from Spain. The sequence in bold in the first row is the published presumed historical human mitochondrial DNA sequence from the first humans that left Africa approximately 100 000 years ago.

Name	Ancestral polymorphic base									
	A	**C**	**T**	**C**	**T**	**G**	**A**	**C**	**A**	**A**
Ahmed	T	A	T	G	T	G	A	C	A	A
Celia	T	C	T	C	T	G	C	C	G	A
Farah	T	A	T	G	T	G	A	C	A	A
Indira	A	C	T	C	T	C	A	T	A	A
Josh	T	C	T	C	T	G	A	C	G	A
Li	A	C	T	C	T	C	T	C	A	A
Lucia	A	C	G	C	T	C	T	C	A	A
Nicky	T	C	T	C	T	G	C	C	G	A
Pablo	T	C	T	C	T	G	A	C	G	A
Sofia	T	C	T	C	T	G	A	C	G	A
Zhang	A	C	T	C	T	C	T	C	A	A

Question 3: Which students have identical DNA sequences at all 10 polymorphic sites? What does this tell you? (10%)

Question 4: Define the term 'haplotype' and calculate how many different mitochondrial DNA haplotypes can be found in your class. (15%)

Question 5: Which haplotypes differ by a single base-pair change? Can you determine within which line the DNA mutation has occurred? (15%)

Question 6: Construct an evolutionary tree that includes all 11 of the students in your class. Clearly show where each DNA base-pair change is likely to have occurred. (25%)

Question 7: Three of the students, Josh, Lucia and Sofia, are from the Americas. What do their mitochondrial DNA sequences reveal about their maternal ethnic origins? (10%)

Chapter 1.6

The Footballer, his Wife, their Kids and her Lover

You've just completed your final school exams, and you're very excited about the prospect of going to university. What you really want to be is a forensic scientist, so your biology teacher has sensibly advised you to study for a genetics degree. You know that students often end up in lots of debt, so you're delighted when you're offered a summer job at your local quality newspaper, the *Daily Globe*. You will be working with Ed Ball, the sports reporter. During the first few days you realise that whilst Ed is hugely enthusiastic and very hard-working, he isn't exactly the cleverest person in the world.

A week later you arrive at work and notice a very animated Ed in a heated 'discussion' with a very smartly dressed and very important looking woman. She, you later discover, is a lawyer. They are inspecting a large parcel with the words 'Private and Confidential' emblazoned on the front.

Once Ed has calmed down, he explains what's in the parcel. It contains an anonymous letter, results of a DNA test and other material. According to the letter, all is not well in the 10-year marriage of Percy Nutmeg (see Fig. 1.6.1), the international fashion icon and top striker with local football team Barnstoneworth United, and his stunningly gorgeous and equally blonde wife Tracy, a moderately talented singer. You were aware they had four sons, Ethan, Jacob, Noah and Oscar, and you're vaguely aware that they've recently had a fifth son, Frank.

Figure 1.6.1 Even from a young age, Percy Nutmeg was a highly successful footballer.

Genetics? No Problem!, First Edition. Kevin O'Dell.
© 2017 John Wiley & Sons Ltd. Published 2017 by John Wiley & Sons Ltd.

The anonymous letter implies that Percy suspects his wife has been 'playing away from home' with Franco Pallone, ace Italian striker from local rivals Denley Moor. Percy is incandescent with rage. However, in a rare and rational moment of calm he has collected toothbrushes from each member of his family and has sent these to ProTest, the leading private DNA profiling enterprise.

It is the DNA profiling data from ProTest that has been sent to you. Ed is totally confused by the data and believes that you, with your interest in genetics and forensics, can interpret the information for him. He is desperate to write an exclusive story, but he wants you, as an aspiring forensic scientist, to make sure he gets his facts right.

You soon realise that the data from ProTest has been supplied as an image of an electrophoretic gel showing variation at a single short tandem repeat (STR). The image is reproduced in Fig. 1.6.2.

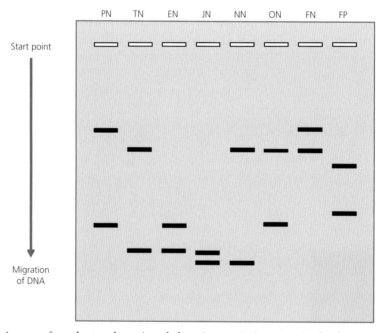

Figure 1.6.2   An image of an electrophoretic gel showing variation at a single short tandem repeat (STR) within the Nutmeg family. Each lane represents DNA from one member of the family. The owner of each DNA sample is indicated with their owner's initials: Percy Nutmeg (PN), Tracy Nutmeg (TN), Ethan Nutmeg (EN), Jacob Nutmeg (JN), Noah Nutmeg (NN), Oscar Nutmeg (ON), Frank Nutmeg (FN) and Franco Pallone (FP).

Ed has little understanding of genetics and has a number of questions. He is naturally concerned about being sued and stresses that as long as your answers are factually correct, he is happy to deal with the lawyers.

Question 1: What is the genetic basis of STRs? (10%)

Question 2: In paternity disputes such as this, why are STR analyses more reliable than those using single nucleotide polymorphisms (SNPs) or other types of DNA variation? (15%)

Question 3: Using a clean and uncontaminated DNA sample as your start point, briefly describe the experimental strategy that culminates in the image of two bands in each lane as shown in Fig. 1.6.2. (20%)

Question 4: According to the data shown in Fig. 1.6.2, who are the parents of baby Frank? (15%)

Question 5: Are there any other issues in the data that Ed might be interested in? (20%)

Question 6: Ed is concerned about the reliability of the data. What further experiments could you do to reassure him? (20%)

Give Peas a Chance

Ever since Bruno, your new stepdad, arrived on the scene your life has become a lot more interesting. You'd not been entirely impressed when you first met him, but had tolerated his strange obsession with all things science fiction because having him around seemed to make your mum much happier. It was of course rather odd that they would occasionally disappear to *Star Trek* conventions together, but at least it meant that from time to time you could have the house to yourself for the weekend.

Sometimes Bruno took his sci-fi obsession a bit too seriously for your liking. He spent a lot of time in his shed at the bottom of the garden (see Fig. 1.7.1), although it wasn't really a shed at all; it was Bruno's 'laboratory'. 'I'm making a time-travelling machine,' he said pointing at something that looked suspiciously like an old portable toilet. 'You use the time portal to key in the date you want to travel to and the name of the person or place you want to visit,' he said pointing at what looked like an old phone glued to the portable toilet door. Or at least that's what you think he said, as by this point you weren't really listening anymore.

Figure 1.7.1   Bruno's shed at the end of the garden. Or is it a laboratory and time portal?

A few weeks later, Bruno and your mother are away attending a *Next Generation* convention and you take advantage of their absence to throw a party for your school friends to celebrate the fact that you've been accepted into university to study genetics. Having consumed your mother's extensive collection of whiskies, your friends leave, but before you go to bed you can't resist the temptation to visit the shed.

In your tired and emotional state you begin to think that Bruno is a little too good to be true. Who is he, where does he come from and does he really know how to build a time machine? You suddenly have a brilliant idea. What about visiting Bruno on the day he was born! Without any great hope or expectation you sit on the portable toilet and press a few buttons. Your hand is shaking as you type 'B R U N O' and, rather than waste time with '8.07pm on 1ˢᵗ August 1962' (Bruno's date of birth) type '20 07 1 8 62'. You look around and expect nothing to happen.

Seconds later the portable toilet takes on a life all of its own and appears to plunge through the darkness. You take a very deep breath and close your eyes, then after what seems like an eternity but is probably less than a minute of real time, everything calms down. You focus and compose yourself before plucking up the courage to open your eyes.

A smiling middle-aged woman, who, as you can see from Fig. 1.7.2, is wearing some very old-fashioned clothes, is waiting to welcome you. 'I'm Mona Sterri-Garten,' she says in what seems to be a strong southern German accent. 'Welcome to Gregor's 40th birthday party. You're just in time as they're about to cut the cake. Richard Bunsen made it. You should see the candles!'

Figure 1.7.2   Mona Sterri-Garten (seated) with her grand-daughter.

Eventually you realise that you must have arrived at Gregor Mendel's 40th birthday celebrations. In fact, as Fig. 1.7.3 shows, Gregor Mendel is wearing his very best party clothes. Incredibly your stepdad's time portal has worked! But you still aren't entirely sure how or why you've arrived in Brno on Sunday 20th July 1862. The party is a great success and the birthday boy entertains his guests with a stunning performance of *Alles was wir sagen ist gebt den Erbsen eine Chance*, a song he'd apparently written himself. You are sure you haven't heard the song before, but the tune sounds strangely familiar.

Figure 1.7.3 Gregor Mendel at his 40th birthday party.

Later that evening you find Mendel sitting with his friends reflecting on the first 40 years of his life. 'It's all a bit of a disaster,' he says. 'I love teaching, but I've failed my teaching exams,' he continues. 'Twice,' Robert Bunsen kindly reminds him.

'But your weather forecasting is the best in the state and your bees make the very best honey,' says his older sister, Veronika, trying to be positive. 'I realise that,' smiles Mendel, 'But I've been trying to breed pea plants that make extra pollen for my bees, and the strangest and weirdest things are happening,' he continues.

You notice some of Mendel's handwritten notes on the table, and try to get a sneaky look at his data. In the summer of 1860, he apparently crossed true-breeding round pea plants to true-breeding wrinkled pea plants and collected the seeds. A year later, when the hybrid plants grew from those seeds, to his great surprise all the F1 plants had round peas. That summer he crossed the F1 plants together, collected the seeds and grew these plants in the summer of 1862 and, as if by magic, some wrinkled pea plants had reappeared! In very large letters were written the word *Kontamination!* Mendel's full set of results are shown in Table 1.7.1.

Table 1.7.1   Number of F2 plants with a wrinkled or round pea phenotype.

Phenotype	Number of plants
round peas	5474
wrinkled peas	1850

Question 1: Try to improve Mendel's mood by explaining why the F2 plants are inherited in the ratios in Table 1.7.1. (10%)

Mendel is fascinated by your explanations of how his pea plants inherit their round and wrinkled pea shapes. He rushes to his room and returns with data from six other pea characteristics, all of which seem to be inherited in a similar fashion.

You ask Mendel whether he's tried crossing peas with the different characteristics together and, much to your delight, he shares his new data with you. It seems that he used the round and wrinkled strains described earlier and a second characteristic associated with pea colour. In this case, a cross between pure-breeding green and pure-breeding yellow plants gave an F1 that were all green and an F2 that segregated in the ratio 3 green to 1 yellow.

When Mendel crossed pure-breeding wrinkled green peas to pure-breeding round yellow peas, all the F1 plants had round green peas.

Mendel then crossed the F1 round green pea plants with each other to give an F2 generation with four classes of phenotype in the ratio 9 round green, 3 wrinkled green, 3 round yellow to 1 wrinkled yellow.

Question 2: Explain to Mendel why crossing F1 round green pea plants to each other gives an F2 generation pea plants in the ratio 9 round green, 3 wrinkled green, 3 round yellow to 1 wrinkled yellow. (15%)

Question 3: Would Mendel get the same results if he'd started his first cross between pure-breeding round green pea plants and pure-breeding wrinkled yellow pea plants, rather than the cross between pure-breeding round yellow and wrinkled green as described earlier? (10%)

You feel you're on a bit of a roll and can't resist the opportunity to explain to Mendel what would happen if he included a third characteristic (plant height, where a cross between tall plants and short plants gives an F1 that is always tall) into his crossing scheme.

Question 4: What classes and ratios of offspring would Mendel expect to see in the F2 if he were to cross a pure-breeding tall round green strain to a pure-breeding short wrinkled yellow strain? (15%)

Mendel is really excited at your explanations, but you sense he doesn't quite believe you. 'But it's not always 3:1,' he says, as he hands you an old parchment. It is some data from the 1830s from Johann Karl Nestler, Mendel's inspirational teacher. Nestler had been investigating the inheritance of black and grey coat colour in Romanian Tsurcana sheep. Apparently Mendel had collected all of Nestler's data together and was very confused. A summary of the Tsurcana sheep data is shown in Table 1.7.2.

Table 1.7.2   The results of crosses between Tsurcana sheep with specific coat colours.

Parents	F1 offspring
black × black	all black
black × grey	50% black, 50% grey
grey × grey	67% grey, 33% black

Question 5: How is coat colour inherited in Tsurcana sheep? (25%)

Question 6: Why might the coat colour gene you've identified in Tsurcana sheep also have a role in another characteristic? (25%)

You suddenly realise Mendel is getting a bit blurry when you hear the unmistakable voice of your mother shouting, 'Have you been asleep all weekend?'

Noah's ARC

It is another scorching hot summer day at Noah's Animal Rescue Centre. You love all the animals and it really is the perfect summer job, except for the smell. You wish there was some way in which you could teach all the fluffy animals about personal hygiene. You have discussed the possibility of some kind of air-conditioning system with the owner, Noah Lott, as he seems to have the answer to everything. 'Global warming,' he says, and that is the end of the conversation.

Your favourite animals at the rescue centre are the guinea pigs, and you are surprised at quite how many different varieties of the little beasts there seem to be. On the rare occasions that Mr Lott allows you a tea break, you have taken some photos of the most photogenic animals. After three summers with Mr Lott at Noah's ARC, your favourite image is still the two guinea pigs, Bubble and Squeak, that you met on your first day (see Fig. 1.8.1).

Figure 1.8.1 Your favourite guinea pigs: Bubble (left) and Squeak.

Genetics? No Problem!, First Edition. Kevin O'Dell.
© 2017 John Wiley & Sons Ltd. Published 2017 by John Wiley & Sons Ltd.

Of course, the main objective of the rescue centre is to rehouse the unwanted animals that arrive there. So all the animals are kept in single-sex groups to avoid 'creating new baby animals' as Mr Lott puts it. However, you decided this was a great opportunity to do a bit of experimentation before you went to university to study genetics. Your first thought was that the combinations of coat colours and patterns shown by Bubble and Squeak were a bit complicated and unlikely to be a good place to start, so you decided to start your investigations with Hazel, a brown female guinea pig from a long line of completely brown animals, and Cloud, a white male guinea pig from a pure-bred strain of brown guinea pigs.

That first summer, you secretly introduced Hazel to Cloud and were delighted when 2 months later, four cream guinea pigs were born. Even better was that because your young niece was so excited by her present of Hazel, Cloud and their pups, she rewarded you with some of her very best drawings in appreciation (see Fig. 1.8.2).

Figure 1.8.2   Your niece's drawings of Cloud (left), Hazel (right) and one of their pups (centre).

Over successive summers you made further crosses between Hazel, Cloud and their offspring. A summary of that data is shown in Table 1.8.1.

Table 1.8.1   Proportion of offspring with specific colour phenotype from multiple crosses between cream, light-brown and white guinea pigs.

Mother	Father	Colours of offspring (%)		
		White	Light-brown	Cream
white	light-brown	0	0	100
white	white	100	0	0
light-brown	light-brown	0	100	0
cream	cream	25	25	50

Question 1: Speculate on how cream, light-brown and white body colour is inherited in guinea pigs. (20%)

Question 2: Using appropriate genetic notation, redraw each of the crosses shown in Table 1.8.1 showing the genotypes of the parents and their offspring. (20%)

To confirm your ideas concerning the way guinea pig body colour is inherited, you decide to carry out some more crosses in secret. Firstly, you decide to cross a light-brown female guinea pig to a white male, and discover that, just like the reciprocal cross from in Table 1.8.1, all the offspring are cream.

Question 3: If you were to cross a cream female to a white male, what ratio of body colours would you expect to see in their offspring? Explain. (10%)

Question 4: If you were to cross a light-brown female to a cream male, what ratio of body colours would you expect to see in their offspring? Explain. (10%)

You discuss your results with your biology teacher, Mrs Cavy. She originally trained as a biochemist and she immediately tells you she thinks the guinea pig body colour gene you've identified encodes an enzyme in a pigment pathway.

Question 5: If Mrs Cavy is correct that the guinea pig body colour gene you've identified encodes an enzyme in a pigment pathway, explain how variation at a single gene can generate three different body colour phenotypes? (20%)

However, when Mrs Cavy accesses the guinea pig genome project and looks at the DNA sequence of the chromosome region where you expect the body colour gene to be, she doesn't find a DNA sequence that looks like it encodes an enzyme from a pigment pathway. You assure her that you've identified the correct chromosomal region and realize there must be another explanation.

Question 6: If the guinea pig body colour gene does not encode an enzyme in a pigment pathway, speculate on what other kind of protein the body colour gene might encode. In addition, explain how might it generate the three body colour phenotypes caused by variation at the body colour gene. (20%)

The Mysterious Disappearance of Midnight

The ear-splitting scream woke you from a very deep sleep. Within moments you were running downstairs in a substantial panic. It was the first time your parents had left you 'home alone' with your younger sister and now she'd broken a leg, or burnt her hand in the toaster, or seen a mouse/spider/snake in the kitchen. It was only when you found your sister, still screaming and panicking in the back garden, that you realised quite how serious things actually were. Your sister was covered in blood.

You tried to calm yourself and work out exactly what had happened. Fortunately, when you were at school you attended a first-aid course as at the time you'd wanted to study medicine and were told you'd need something more than just a fabulous set of exam results. You carefully cleaned all the blood from your sister's arms and began to realise that she wasn't actually injured after all. So where had all the blood come from? Nine-year-old sisters are invariably a little odd and annoying, but very few are involved in gratuitous slaughter.

Eventually your sister calmed down just enough to begin to explain what had happened. 'It's Midnight,' she said, 'Midnight'. You glanced at the clock. It was half past one. 'It's half past one,' you replied helpfully, pointing at the clock. 'No it's Midnight. Midnight,' she repeated somewhat more hysterically. The horrible truth suddenly dawned on you and there was a sick feeling in the pit of your stomach. Midnight was the name of your sister's much-loved pet black rabbit (see Fig. 1.9.1). You grabbed a torch and ran to Midnight's hutch. Outside you found the blood-stained body of Midnight. You put Midnight in a brown paper bag and immediately phoned the police.

Genetics? No Problem!, First Edition. Kevin O'Dell.
© 2017 John Wiley & Sons Ltd. Published 2017 by John Wiley & Sons Ltd.

Figure 1.9.1   Your sister's picture of Midnight in happier times.

Question 1: If you want to retain the body of a recently killed animal for forensic analysis, why is putting it in a paper bag, rather than a plastic bag, a sensible idea? (10%)

Thirty minutes later there was a knock at the door and policewoman Sheila Holmes entered the house, along with Dr Jane Watson, a forensic scientist who specialises in crimes involving animals. You explained to your sister that Dr Watson needed to take Midnight away to help her find out who hurt him. You also told Dr Watson that you are soon to start a genetics degree and would be really interested in helping out if at all possible. Dr Watson suggested you came to her laboratory later that morning. You cleared up the mess in the house, put your sister to bed and tried to get some sleep.

Late the following morning you arrived at Dr Watson's forensics laboratory. She'd been very busy reviewing the CCTV footage from the previous evening and found seven dogs that we're nearby when Midnight was attacked. Three of the dogs are rottweilers belonging to Bob and Peter McNasty, collectively known as the Notorious Boys. They are well known to the police. Dr Watson has also seen three greyhounds that are owned by three local enthusiasts of greyhound racing, and a corgi that belongs to an old lady from the Palatial Village Retirement Home. Dr Watson also identified two foxes that were seen in the vicinity. The nine animals are shown in Fig. 1.9.2 and their details, as known to policewoman Sheila Holmes and Dr Watson, are listed in Table 1.9.1.

Figure 1.9.2 The suspects: (a) Fang, (b) Slayer, (C) Diablo, (d) Bolt & Lightening, (e) Duchess, (f) Flash, and (g & h) the foxes.

Table 1.9.1 The nine animals tested, including their apparent breed and alleged owner. The code number links to the DNA samples shown in Fig. 1.9.1. Clearly the foxes neither have a name nor an owner.

Code	Dog name	Apparent breed	Alleged owner
1	Fang	rottweiler	Bob McNasty
2	Slayer	rottweiler	Bob McNasty
3	Diablo	rottweiler	Peter McNasty
4	Bolt	greyhound	Charles Munn
5	Lightening	greyhound	Brigadier-General Critchley
6	Duchess	corgi	Elizabeth Queen
7	Flash	greyhound	Major Lyne-Dixon
8		female fox	
9		male fox	

You were very impressed that Dr Watson had already obtained DNA samples from all nine of the suspect animals. She had made DNA samples from Midnight's wounds, as well as from saliva on Midnight's coat.

Question 2: Why should the saliva sample give a better DNA profile than the DNA sample from the wound? (10%)

Dr Watson explained that she has developed a series of species-specific DNA tests that not only identifies the species from which a DNA sample originates, but also distinguishes between, and therefore identifies, individuals within that species. She explained that she uses a polymerase chain reaction (PCR) test that amplifies short tandem repeats (STRs).

Question 3: Briefly explain how PCR can be used to amplify STRs. (10%)

Question 4: How can STRs be used to identify a specific species? (10%)

Question 5: How can STRs be used to distinguish between specific individuals within a species? (5%)

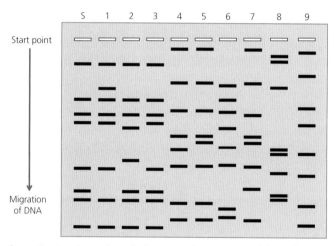

Figure 1.9.3   Image of an electrophoretic gel showing four pairs of STR fragments from in the saliva sample found on Midnight (S), as well as equivalent samples from the seven dogs (1-7) and two foxes (8-9) described in Table 1.9.1.

Question 6: Using the DNA evidence in Fig. 1.9.3, explain which animal's saliva has been found on the body of Midnight? (5%)

Question 7: What statement should Dr Watson make to PC Holmes regarding which animal (or animals) attacked Midnight? (10%)

Dr Watson allowed you to make a copy of the data shown in Fig. 1.9.3 to take home with you. 'There's quite a lot of interesting information here,' she said. 'Have a read of my book and see what else you can find out.' You read Dr Watson's book *Forensic Science* with great enthusiasm. You discover that human forensics is based on 13 pairs of STRs. They are inherited in a standard Mendelian fashion, so you get 13 from your mother and 13 from your father. It follows that you have about 6 or 7 of these STRs in common with each of your grandparents. Because of this, STRs are used extensively in paternity tests and other situations where a close genetic relationship is suspected. Apparently, two people taken at random normally have no, or very few, STRs in common.

Question 8: What would be the genetic relationship between two people if they shared all 26 STRs? (5%)

Having read Dr Watson's fascinating book from cover to cover, you looked again at the data in Fig. 1.9.2. You were not entirely convinced that the data made a lot of sense, particularly that from the dogs. You discussed your concerns with Dr Watson, and she suggested you try to answer the following questions.

Question 9: How many STRs do the two foxes (8 & 9) have in common? (5%)

Question 10: What, if any, is the genetic relationship between the two foxes? (5%)

Question 11: How many STRs do the three rottweilers (1, 2 & 3) have in common? (5%)

Question 12: How many STRs do the three greyhounds (4, 5 & 7) have in common? (5%)

Question 13: Speculate on why the nature of STR variation in dogs is so different from that of foxes and humans. (15%)

Chapter 1.10

RANCID

It wasn't the summer job you'd hoped for. You'd been working hard during the year and not only had you passed the exams that would allow you to start a genetics degree, you'd actually won your school prize! You'd contacted lots of research labs hoping they'd let you work with them for the summer but hardly anyone replied and nobody had a job for you.

So you return to the local cat rescue centre where you'd spent the previous couple of summers. Rescue All Native Cats in Danger (RANCID), to give the rescue centre its proper name, looks after hundreds of unwanted cats and attempts to rehouse them (see Fig. 1.10.1). However, it really is the summer job from hell. It isn't the scratches and bites from the thoroughly ungrateful animals that annoys you the most. It's the smell. There are hundreds of cats, they continually mark their complex territories and they produce poo on a truly industrial scale. You won't make the mistake of wearing any decent clothes to work this time, as you know you'll have to burn them at the end of the summer.

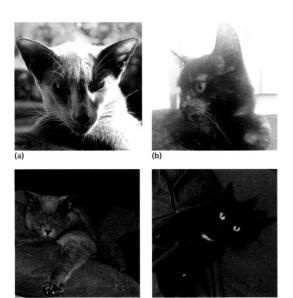

Figure 1.10.1   Some of the new arrivals at RANCID.

Genetics? No Problem!, First Edition. Kevin O'Dell.
© 2017 John Wiley & Sons Ltd. Published 2017 by John Wiley & Sons Ltd.

The person in charge of RANCID is Selina Kyle, who you and the other employees usually refer to as Catwoman, though never to her face. Interestingly she has just become aware that you're about to embark upon a genetics degree, so this summer she has a special project for you. Apparently a lot of Selina's customers are interested in acquiring cats for breeding projects and she sees a great marketing opportunity. She wants you to further develop her cat database and, wherever possible, include details of the genetic make-up of her cats.

You remember some lessons from your old school teacher, Felix Le Chat, and appreciate that cat coat colour, like most characteristics in most organisms, is complex, involving variation at many genes. So you spend much of your first week online trying to get some basic understanding of the genetic basis of coat colour in cats. Selina, who is obsessed by everything cat, provides you with a huge database of cat coat colour that she's compiled. It's a truly impressive document and also contains details of which cats have produced kittens together. You decide to focus your initial studies on Lucky, Selina's favourite cat. Lucky is a male who is completely black and whose image is shown in Fig. 1.10.2. He is also, according to Selina's data, the father of an astonishing number of kittens.

Figure 1.10.2   Images of the three cats from the cat sanctuary, Lucky, Snowy and Chocolate, as well as Meaow who belongs to your teacher Felix Le Chat. The images were drawn by Selina Kyle's daughter.

Lucky first produced kittens with Snowy, the white female drawn in Fig. 1.10.2. To Selina's surprise all 10 kittens they produced were entirely black. The following season some of the siblings mated to produce a grand total of 46 kittens, 34 of which had black coats just like their grandad Lucky, whilst 12 were white just like their grandma Snowy.

Question 1: Speculate on how the black/white body colour might be inherited. (10%)

Question 2: If the body colour gene encodes an enzyme, speculate on how it may cause the body colour phenotype. (10%)

Looking through Selina's database you find a similar pattern of inheritance when Lucky mated with Chocolate, the brown female cat shown in Fig. 1.10.2. All their offspring had black coat colour, and when they mated with each-other they produced kittens in a ratio of three blacks to one brown.

Question 3: Speculate on how the black/brown colour is inherited? (10%)

Selina had obviously been a bit confused at this stage and had mated two of Lucky's 'grandkittens' together. These were a white-coated female and a brown-coated male. This time all their F1 offspring had brown coats just like their father. When the F1 cats were mated together, the kittens produced were in a ratio of approximately 3 brown-coated cats to 1 white-coated cat.

Question 4: What is the genetic basis of the three coat colours: black, brown and white? (10%)

Question 5: In light of your answer to question 4, speculate on the type of mutations that may cause the phenotypes caused by the mutant alleles and explain their effect on the enzyme they encode. In addition, explain why you see the dominance/recessive relationships described earlier. (20%)

You're rather excited by the results you've discovered and show them to your old school-teacher Felix Le Chat. 'Very well done,' he says. 'You should get hold of a Siamese cat,' he continues. 'You might find the results a whole lot more interesting!'

Felix hands you a picture of his Siamese cat Meaow, who is reproduced in Fig. 1.10.2. She has an almost white body colour, except at the extremities where she is very black.

Question 6: How might a mutation in a gene that encodes an enzyme lead to the phenotypes you see in Meaow? (10%)

Question 7: If Meaow, or her Siamese brother, had kittens with Lucky, Chocolate or Snowy, what would the phenotype of each of the three sets of kittens be? (20%)

You undertake further studies of the four cats and discover that Snowy is profoundly deaf, but the other three cats have normal hearing. You read about this online and find that white-pigmented mammals are often hearing impaired.

Question 8: Speculate on why Snowy, and other white-pigmented mammals, might be hearing impaired. (10%)

Intermediate

Chapter 2.1

Otto's Finger

According to your aunt, it was entirely the outcome she'd predicted. 'Don't go on a skiing holiday until you've finished your degree,' she'd said. 'If you break your leg you'll regret it.' And you did. And you do. However, your aunt doesn't know the whole story and it could be the luckiest break you've ever had.

It had seemed a particularly good idea at the time, having a New Year skiing trip with friends from the Genetics course at university, especially as you were visiting the fabulous Swiss Canton of Valais (see Fig. 2.1.1). Obviously you wouldn't have time to go away later in your final year, as you'd be studying. Or at least that was the plan. In fact, the first five days of your holiday had been fantastic and your skiing was coming on in leaps and bounds. Then someone suggested going off-piste (or at least you think that's what they said) and it all went horribly wrong. The last thing you remember is the spectacular view of the Otemma Glacier (see Fig. 2.1.2), all be it at a rather peculiar angle.

Figure 2.1.1 The Swiss Alps.

Genetics? No Problem!, First Edition. Kevin O'Dell.
© 2017 John Wiley & Sons Ltd. Published 2017 by John Wiley & Sons Ltd.

Apparently you were only unconscious for about 15 minutes and, as you slowly opened your eyes, you saw a look of total horror on the faces of your friends. You felt a bit dazed and confused, but assumed everything was fine as you seemed to have landed in a huge pile of snow next to a mountain stream. Your relief was short-lived, however, as you suddenly realised your leg was at an angle that could best be described as somewhere between unnatural and impossible. And there was an awful lot of pain.

Figure 2.1.2   The Otemma Glacier.

You were beginning to think that your friends were completely and utterly useless, as they'd done little or nothing to help. Then you realised they were pointing at something behind you. You slowly turned your head, expecting to see a rock coated in bits of your leg, only to have the biggest surprise of your life. There, in the snow next to you, was a very hairy man and he was most surely and definitely, dead. You were suddenly gripped by a feeling of panic. Surely you hadn't landed on him and killed him?

You regained consciousness for the second time in Martigny Regional Hospital. Your leg was now in the traditional position, at an appropriate angle and contained within a cast. Thankfully the intense pain had disappeared. Your friends were nowhere to be seen, but there were two people in your room, a nurse and an extremely serious looking man in the uniform of the Valais Police. Over the next half an hour you discovered how global warming was apparently causing the Otemma Glacier to melt into the stream you'd landed next to. Fortunately, as the body you'd landed on was probably thousands of years old and until recently had been frozen within the glacier, this meant you definitely hadn't killed anyone. 'Your friends tell me you're a genetics student,' continued the policeman, 'and you're very interested in the evolution and migration of human populations in Europe? Would you like to take a small part of the iceman home with you for your studies?'

Once back at university, you immediately contact the head of the Human Evolutionary Research Institute (HER Institute), the independently minded Professor Lauren Towers-Elf. To your great relief, she was delighted you'd brought an ancient frozen finger from Otto's left hand to visit her. Having limped carefully into her office, you begin to explain how you had managed to find the Otemma Glacier Iceman. Professor Towers-Elf has a rather intimidating reputation as one of the greatest intellectuals of your time and you expect her to make a particularly profound statement about the significance of the Otemma Glacier Iceman's finger. 'I think we should call him Otto,' she says.

Professor Towers-Elf kindly allows you to work for nothing in her laboratory. Her first suggestion is that you try to get a mitochondrial DNA profile from Otto.

Question 1: As Otto was found in Europe, he must have died within the past 40 000 years. How could you determine more precisely when he died? (5%)

Question 2: Why do studies of ancient remains often focus, at least initially, on mitochondrial DNA (which is precisely 16 569 bp long) rather than nuclear DNA (which is approximately 3 billion base pairs long)? (10%)

On hearing of your 'accidental' success in acquiring a research placement in the laboratory of Professor Towers-Elf, your aunt is suddenly interested in a genetic analysis of her, and therefore your, family tree and sends you a copy of her genealogical research effort. The document is quite impressive in its detail. Indeed, it contains complete and apparently accurate information on the previous six generations of your ancestors.

Question 3: From how many different people who were alive six generations ago have you acquired (a) your nuclear DNA and (b) your mitochondrial DNA? Are there any circumstances under which these theoretical estimates could be larger or smaller? (10%)

Having acquired a full mitochondrial DNA sequence from Otto's finger, you compare his 16 569 bp mtDNA sequence to the many tens of thousands of sequences in the vast Mitochondrial Eve Project database. Initial indications are that Otto belongs to haplogroup K, a mitochondrial haplogroup prevalent in Eurasia, North Africa and South Asia that is thought to have arisen about 12 000 years ago. By a remarkable stroke of luck, Professor Towers-Elf has just initiated a study of haplogroup K and that very morning has 18 new haplogroup K mitochondrial sequences to analyse (see Fig. 2.1.3).

Figure 2.1.3 Liquid Nitrogen Facility in HER institute where Otto's finger and the remains of other ancient humans are stored under perfect laboratory conditions.

Professor Towers-Elf asks you to establish the evolutionary relationship between Otto and the 18 new sequences. She also suggests that you compare their mitochondrial DNA sequences to that of Oetzi, the 5300-year-old Iceman found on the Austrian–Italian border in 1991. These data, presented as a list of mitochondrial nucleotide differences between the progenitor K1 haplogroup and each of the 20 new mitochondrial DNA sequences, are shown in Table 2.1.1.

Table 2.1.1 Nucleotide differences between the mtDNA sequences of twenty K haplogroup individuals relative to the established definitive historical progenitor K1 haplogroup. The position of the affected nucleotide is shown. Note that within this analysis any nucleotide change (mutation) arises only once within the lineage and is therefore a unique mutational event.

Name	Nucleotide changes from K1
Adelina	1819, 5090, 6413, 8251, 10478, 12810, 14142, 16093, 16274, 16362
Aldric	498, 3738, 9006, 14002, 14040, 16320
Baldomar	5913, 7521, 8709, 12738, 12771
Berhta	709, 1189, 4561, 9716, 10398, 14305
Bruno	1819, 5090, 6413, 8251, 10478, 12810, 14142, 16093, 16274, 16362
Chlothar	497, 3394, 5093, 6260, 11485, 11840, 13740
Egilhard	497, 13117
Ermingard	709, 1189, 4561, 8618, 9716, 10398
Frida	497, 6260, 11485, 11840, 13740, 16311
Gasto	497, 7118, 11485, 12017, 12399
Hengist	1819, 5090, 6413, 8251, 10478, 12810, 14142, 16093, 16274, 16362
Hrolf	709, 1189, 4561, 9716, 10398, 11549
Odila	498, 9006, 12358, 14002, 14040, 14506, 15244, 16320
Oetzi	3513, 8137, 16362
Otto	3243, 8137, 16362
Radulf	5913, 7521, 8709, 12738, 12771
Rosmunda	5237, 5913, 6845, 10154, 12738, 15301
Saxa	498, 9006, 12358, 14002, 14040, 14506, 15244, 16320
Tancred	498, 508, 9093, 11377
Wido	709, 1189, 4561, 9716, 10398, 16189

Question 4: Use the data in Table 2.1.1 to construct an evolutionary tree showing the relationship between all 20 individuals. (20%)

Question 5: Is the evolutionary tree you have constructed robust and reliable? How could you improve its reliability? (5%)

You look more closely for any functional consequences of changes in Otto's mitochondrial DNA sequence and focus on his A3243G mutation. This mutation is within the mitochondrial tRNA(Leu) gene and is well established as a cause of a condition called mitochondrial encephalomyopathy, lactic acidosis and stroke-like episodes (MELAS). According to the extensive literature on the subject, MELAS primarily affects the brain (encephalo-) and muscles (myopathy) and symptoms may include muscle weakness and pain, recurrent headaches, loss of appetite, vomiting and seizures. The pedigree of a typical MELAS family is shown in Fig. 2.1.4.

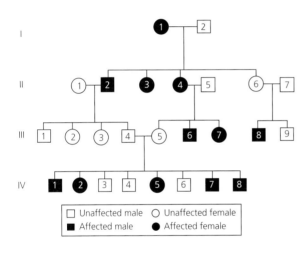

Figure 2.1.4 Pedigree of a typical family in which the MELAS A3243G mitochondrial mutation is prevalent.

Question 6: What is the evidence that the MELAS family is suffering from a mitochondrial condition? (10%)

Question 7: Why do mutations in the mitochondrial genome cause such variable phenotypes and in some cases even seem to skip generations? (10%)

Question 8: Speculate on the functional consequence of a mutation in a mitochondrial tRNA(Leu) gene? Why should it lead to such a broad range of phenotypes affecting so many organs and systems in the body? (10%)

MELAS frequently appears in childhood, often after a period of apparently normal development. However, it can be a late onset condition. For example, it's been speculated that Charles Darwin (see Fig. 2.1.5) may have been suffering from MELAS, although other sources suggest Crohn's disease. In his autobiography Darwin wrote 'Even ill-health, though it has annihilated several years of my life, has saved me from the distractions of society and amusement'.

Not unreasonably, you consider that managing to acquire a full mtDNA sequence from a 5000-year-old man is quite an achievement. However, Professor Towers-Elf is not so easily satisfied and rather strongly encourages you to acquire some nuclear DNA sequence from Otto's finger. You acquire a partial DNA sequence that includes the entire lactase (*LCT*) gene sequence from chromosome 4.

You know that the *LCT* gene encodes the enzyme lactase, which breaks down the disaccharide sugar lactose into the monosaccharides glucose and galactose. Lactose is the most abundant sugar in milk. Genetic analyses have revealed that there are two common versions of the *LCT* gene in European populations. The first Europeans were lactose intolerant while the majority of modern Europeans are lactose tolerant. From a genetic perspective, the difference in phenotype is associated with two single nucleotide polymorphisms (SNPs) situated several thousand bases pairs upstream of the *LCT* gene. A lactose intolerant individual has the ancestral haplotype C-13910 & G-22018 (where -13910 means 13910 bases upstream of the ATG start codon of the LCT gene). A lactose tolerant European has the derived haplotype T-13910 & A-22018. Your sequencing data reveals that Otto is homozygous for the C-13910 & G-22018 haplotype.

Figure 2.1.5   Charles Darwin. Did he suffer from MELAS?

Question 9: Propose a mechanism by which the C-13910T and G-22018A mutations in the human *LCT* gene might lead to a change in human phenotype from lactose intolerant to lactose tolerant. (10%)

Question 10: Otto is homozygous for the lactose intolerant haplotype, but some of his friends may well have been heterozygous. Would Otto's heterozygous friends have been lactose tolerant or intolerant or intermediate? Explain why. (10%)

Chapter 2.2

The Mystery of Muckle Morag

You are a genetics undergraduate student and you realise that your final year is fast approaching. So, for your last summer of freedom, you decide to embark upon a long summer holiday before the hard work of your final year begins. You elect to spend the summer on the small picturesque, but very isolated, Scottish island of Muckle Morag (see Fig. 2.2.1).

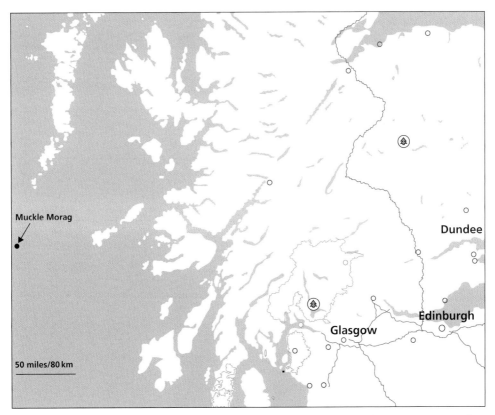

Figure 2.2.1   Map showing the position of the Island of Muckle Morag off the west coast of Scotland.

Genetics? No Problem!, First Edition. Kevin O'Dell.
© 2017 John Wiley & Sons Ltd. Published 2017 by John Wiley & Sons Ltd.

You know very little about the history of Muckle Morag, but on your first night in Ned's Bar, the only pub on the island (see Fig. 2.2.2), you are introduced to Mad Jack, the oldest and wisest man on the island. Over several hours he fascinates you with the Mystery of Muckle Morag.

Figure 2.2.2 Ned's Bar on the Island of Muckle Morag. The image shows quite clearly why the local population jokingly refer to it as 'Ned's Barn'.

According to Mad Jack, the first humans settled on Muckle Morag on 26th November 1703. A young couple, Alex and Morag, and their young son Cameron were in their fishing boat *The Griselda* (see Fig. 2.2.3), when they encountered The Great Storm. *The Griselda* was completely destroyed, but by some miracle and the awesome swimming skills of Morag, the family reached the uninhabited (and at that point unnamed) island of Muckle Morag.

Figure 2.2.3 A modern reproduction of *The Griselda* as constructed by the Muckle Morag Historical Society.

Remarkably, the family thrived in glorious isolation for over 200 years and when they were rediscovered, they comprised an apparently healthy population of 50 individuals. Whilst in recent history, a few people have moved to Muckle Morag, nearly everyone on the island claims to be descendants of Alex and Morag. However, continued Mad Jack, legend has it that in the Great Storm young Cameron was struck in the face by lightning and as a result had a very small and misshapen nose. You look around the bar and suddenly realise that many of the inhabitants have small, misshapen noses.

Curiously, Mad Jack has a perfectly normal looking nose but as he is getting tired and emotional you decide not to point this out. You make your excuses and leave, but you can't get the Mystery of Muckle Morag out of your mind. Through your extensive training as a genetics undergraduate you realise there must be a rational scientific explanation for the 'Cameron' nose and begin to wonder whether the condition is in fact inherited.

Over the rest of the summer you covertly research the family history of the islanders. To help you in this quest you also secretly collect forensic samples (such as spit, hair follicles and, in one case, a significant volume of blood) and store them in the fridge in your holiday cottage. These data allow you to construct the genetically accurate family tree that is presented in Fig. 2.2.4.

On your return to university in September, you realise you are going to do the best final year genetics research project ever!

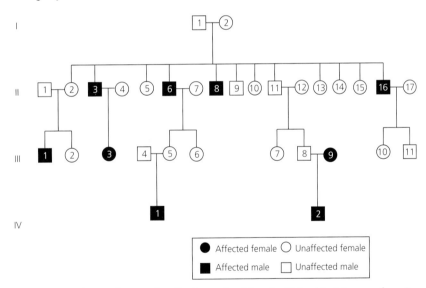

Figure 2.2.4   Extended pedigree from a family from the Island of Muckle Morag, showing prevalence of the 'Cameron' nose (dark squares and circles). Mad Jack is individual I:1.

Question 1: Discuss the evidence supporting the idea that the 'Cameron' nose is caused by an X-linked recessive condition? (5%)

You decide to call the putative small misshapen nose gene *cameron* and use the symbol *cam*.

Question 2: Redraw the family tree showing, where possible and using appropriate notation, the *cameron* genotypes of all the members of the family. (10%)

Having established that the *cameron* mutation is on the X-chromosome, you sensibly decide to determine the precise map position of the *cameron* gene within the X-chromosome. However, you haven't got enough DNA from some individuals for a full genome sequence analysis, so you decide to map *cameron* relative to a well-characterised short tandem repeat (STR). The STR is a trinucleotide CAT repeat that varies between 6 and 15 repeats. You add the CAT STR data to the extended pedigree from the family from the Island of Muckle Morag shown earlier and these data are presented in Fig. 2.2.5.

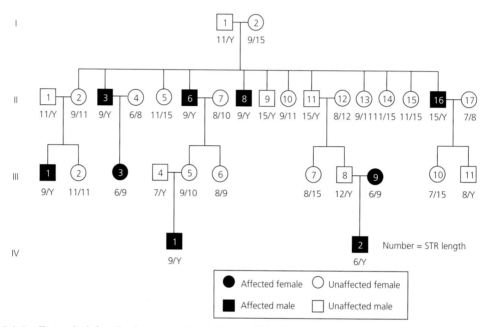

Figure 2.2.5   Extended family from Muckle Morag. This figure is identical to Fig. 2.2.1 except for the addition of STR repeat numbers. An individual with 6/9 has one CAT allele of six repeats and another CAT allele of nine repeats.

Question 3: What evidence is there that *cameron* and the CAT STR are genetically linked? (10%)

Question 4: Calculate the genetic map distance between *cameron* and the CAT STR. (20%)

Question 5: If woman II:2 has a child with her younger brother II:11, what is the probability that they will have a Cameron-nosed son of CAT STR 11/Y? (5%)

Having established the approximate map position of the *cameron* mutation, you contact the editor of the *Muckle Morag Herald* and explain that small, misshapen noses are a genetic condition. In the ensuing excitement, many of the islanders risk serious injury to send you further samples of their DNA to aid your research. They also raise enough money to use next generation sequencing on six affected and six unaffected males.

Question 6: Explain how you would use the 12 whole genome sequences, the family tree, the mapping studies and your knowledge of human mutations to identify the mutant DNA sequence of the *cameron* gene. What problems might you encounter? (20%)

Your preliminary analysis suggests that the *cameron* gene encodes a cell-surface, membrane-bound receptor. The mutant small, misshapen Cameron nose phenotype seems to be caused by a missense mutation close to the 5′ end of the coding region. In affected males, the levels of *cameron* mRNA and Cameron protein are similar to those of unaffected males. In addition, in unaffected individuals most of the Cameron protein is found in the cell membrane, whereas in affected individuals all of the Cameron protein is found in the cytoplasm.

Question 7: In light of your preliminary analysis, speculate on how the recessive missense mutation might cause the *cameron* mutant phenotype. (10%)

Curiously, the *cameron* gene has already been identified. Unlike the curiously misshapen Cameron nose, the Hooter family from Yorkshire, have very large noses that seem to be caused by a dominant missense mutation in the *cameron* gene (see Fig. 2.2.6). The Hooter mutation in the *cameron* gene does not appear to affect Cameron protein or *cam* mRNA levels.

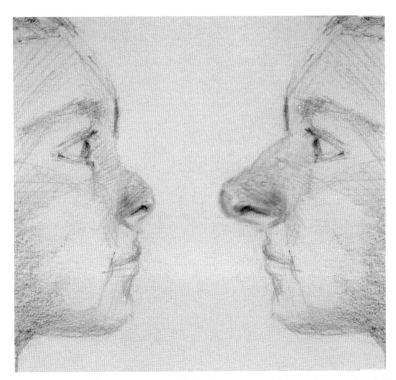

Figure 2.2.6   The Cameron (left) and Hooter (right) noses as drawn by medical illustrator Leanne Ardodavinche.

Question 8: Speculate on how a dominant missense mutation in the *cameron* gene of the Hooter family might result in large noses. (10%)

Question 9: How would you test the model you proposed that speculating on how the dominant missense mutation in the *cameron* gene of the Hooter family causes the large nose phenotype? (10%)

Chapter 2.3

Drosophila hogwashii

As you approach the final year of your genetics degree you are beginning to think about your future career plans. Whilst you thoroughly enjoy most aspects of the subject, you don't really see your future as a full-time genetics researcher. What you really want to be is a science teacher specialising in biology.

Figure 2.3.1 View of Hogwash School showing the staircase leading to the student research laboratories.

You are therefore absolutely delighted when you are given the chance of a year's placement as a trainee teacher in the very prestigious Hogwash School (see Fig. 2.3.1). You are determined to make good use of this opportunity and are keen to ensure your students find genetics as exciting as you do. You suspect the key way to do this is to help your students to undertake their own research.

Genetics? No Problem!, First Edition. Kevin O'Dell.
© 2017 John Wiley & Sons Ltd. Published 2017 by John Wiley & Sons Ltd.

As part of their final school year research project, three of your favourite students, Harriet, Rhona and Hermes, mutagenise a population of wild-type, green-eyed fruit flies of the species *Drosophila hogwashii* (see Fig. 2.3.2) using the mutagen ethyl methanesulphonate (EMS). Whilst they are sorting flies of the F2 generation, they notice a single male fly with puffy yellow eyes. They name this scary looking mutant *vol-au-vent* (*vol*) and decide to study it further.

Figure 2.3.2   A healthy stock bottle of *Drosophila hogwashii*.

Question 1: What affect does EMS have on DNA? (5%)

You explain to your students that *Drosophila hogwashii* has similar chromosomal organisation to the more frequently studied *Drosophila melanogaster*, as it has an XX/XY sex-determination system and three autosomes (non-sex chromosomes) designated chromosomes 2, 3 and 4.

Question 2: How could the students determine whether the *vol-au-vent* mutation is dominant or recessive? (5%)

Question 3: How could the students determine whether the *vol-au-vent* mutation is X-linked or on chromosome 2 or 3 or 4? (10%)

Initial investigations suggest that *vol-au-vent* is a recessive mutation on the *Drosophila hogwashii* second chromosome.

The students are keen to identify the specific DNA sequence of the *vol-au-vent* gene because they would like to know more about the protein that the *vol-au-vent* gene encodes. So they decide to undertake a three-point test cross to determine its position on chromosome 2 more precisely.

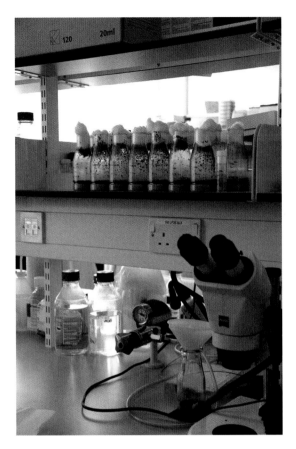

Figure 2.3.3   The well-equipped Hogwash School Fly Room.

Fortunately the school has a well-equipped fly room in which the students can undertake their work (see Fig. 2.3.3). They cross true-breeding yellow-eyed *vol-au-vent* mutant females to true-breeding males carrying two previously described second chromosome mutations called *haggard* (*hag* – which has extra bristles) and *shortbottom* (*sho* – which has seven abdominal segments rather than the normal eight). The resulting F1 females and males are heterozygous for all three mutations and have a wild-type phenotype. The students perform a test cross between a triply heterozygous F1 female and a triply homozygous mutant male. The results of the test cross are shown in Table 2.3.1.

Question 4: Using appropriate notation and showing full genotypes, draw the students' crossing scheme. (10%)

Table 2.3.1   Offspring from a test cross between triply heterozygous F1 females and a triply homozygous mutant males as described in the text. Note that the order of the three genes (left to right) in the table does not necessary imply that this is the order of the three genes (left to right) on the chromosome, because at this stage of the research the order of the genes along the chromosome is unknown.

vol-au-vent	few bristles	eight segments	394
green-eye	*haggard*	*shortbottom*	411
vol-au-vent	*haggard*	eight segments	35
green-eye	few bristles	*shortbottom*	29
vol-au-vent	*haggard*	*shortbottom*	2
green-eye	few bristles	eight segments	3
vol-au-vent	few bristles	*shortbottom*	61
green-eye	*haggard*	eight segments	65
			1000

Question 5: Use the data in Table 2.3.1 to draw a chromosome map that clearly shows the position of the *vol-au-vent* gene relative to the *haggard* and *shortbottom* genes. (15%)

Question 6: Why don't the map distances add up precisely? (5%)

A three-point test cross only gives approximate map positions, so on your advice the students decide to map the mutation more precisely using a set of deletion mutations from the *vol-au-vent* region. For reasons that are entirely unclear, the four deletions are called leo, meles, aquila and vipera. The extent and relative positions of the four overlapping and recessive lethal deletions are shown in Fig. 2.3.4.

Flies heterozygous for the *vol-au-vent* mutation and the leo deletion have yellow eyes, as do flies heterozygous for the *vol-au-vent* mutation and the vipera deletion. However, flies heterozygous for the *vol-au-vent* mutation and the meles deletion have green eyes, as do flies heterozygous for the *vol-au-vent* mutation and the aquila deletion.

Figure 2.3.4   The chromosomal organization of four deletions in the *vol-au-vent* region of chromosome 2. The white area shows the region of the chromosome that is missing (deleted) from the chromosome, whilst the black area shows the chromosomal regions that are present.

Question 7: Redraw the deletion map from Fig. 2.3.4 to show approximately where the *vol-au-vent* gene lies. (10%)

Question 8: Draw a crossing scheme for the deletion mapping, clearly showing the frequency of every genotype and phenotype in each generation when (a) you start with a homozygous *vol-au-vent* female and cross her to a male carrying the leo deletion, and (b) you start with a homozygous *vol-au-vent* female and cross her to a male carrying the aquila deletion. (10%)

Figure 2.3.5 The Pumpkin Pi Logo.

The students use Harriet's Pumpkin Pi 2.0 laptop (see Fig. 2.3.5) to enter the *Drosophila hogwashii* Genome Project Database and look for candidate genes in the *vol-au-vent* region. They find a previously uncharacterized gene which has a sequence that suggests it encodes an enzyme on a pigment biosynthetic pathway.

Question 9: Briefly describe how the students could prove that the pigment enzyme gene is *vol-au-vent*? (5%)

The students continue their experiments and find a misshapen blue-eye colour mutant that they call *malformed* (*mal*). The *malformed* mutation was discovered in a screen for *P*-element-induced eye-colour mutations. *P*-elements are transposons that under certain circumstances can move from one position in the *Drosophila hogwashii* genome to another. If they move to a new chromosomal site they may interrupt the endogenous DNA sequence, which may result in a new mutant phenotype.

Question 10: What are the relative advantages and disadvantages of EMS mutagenesis as against *P*-element (transposon) mutagenesis? (5%)

The students show that the *malformed* mutation is recessive and maps to the third chromosome. Like *vol-au-vent*, the *malformed* gene seems to encode an enzyme from a pigment biosynthetic pathway.

Hermes, who is the brightest and hardest working of the three students, suggests that both *vol-au-vent* and *malformed* encode enzymes in the same pigment biosynthetic pathway. He decides to test this by crossing the mutant strains to generate *vol-au-vent*; *malformed* double mutants.

Question 11: Speculate on what eye colour the *vol-au-vent*; *malformed* double mutant would have if Hermes' theory is correct? (10%)

Question 12: In fact, *vol-au-vent*; *malformed* double mutant flies have white eyes. Speculate on why this is. (10%)

Chapter 2.4

The Curse of Lilyrot

It is the summer before the final year of your genetics degree. You've tried hard to get a job in a high-quality research laboratory, but your grades aren't fantastic, and summer scholarships are very competitive. It looks as though you'll have to work for your aunt again.

As you can see from Fig. 2.4.1, Aunt Tarragon's Garden Centre is a vast enterprise. She is obviously a very astute businesswoman. She also rates herself as a bit of an amateur geneticist, which can be slightly annoying. When you contact Aunt Tarragon about the possibility of a summer job, she is absolutely delighted and tells you she has a very special project for her 'little geneticist'. You approach the summer with even more trepidation than usual.

Figure 2.4.1 Aunt Tarragon's Garden Centre, showing a rather impressive array of garden plants.

Genetics? No Problem!, First Edition. Kevin O'Dell.
© 2017 John Wiley & Sons Ltd. Published 2017 by John Wiley & Sons Ltd.

When you arrive for your first day at Aunt Tarragon's Garden Centre, you are pleasantly surprised that she plants you in front of a top-of-the-range computer on a brand new desk in her office. In previous summers she's only used you to keep the Garden Centre clean and tidy. As the computer springs into life, Aunt Tarragon explains that over the past few years she's been generating her own varieties of lily by crossing various strains together. To your astonishment she's managed to generate a new series of colour variants, many of which are proving extremely popular with her customers. The only snag is that some of them are very susceptible to the viral infection lilyrot. Aunt Tarragon has kept copious notes of the crossing schemes she's been using and asks you to review her data.

At the start of her work, Aunt Tarragon acquired three true-breeding strains of lily. The three independently derived strains were easily distinguished because they each had a particularly distinct flower colour (blue, red or yellow: see Fig. 2.4.2). Aunt Tarragon's data show that breeding two blue-flowered lilies always generated blue-flowered offspring, crosses between two red-flowered lilies always yielded red-flowered offspring, and mating two yellow-flowered lilies always gave yellow-flowered offspring.

Figure 2.4.2 The most popular lily from Aunt Tarragon's Garden Centre.

Question 1: From a genetic perspective, what is the significance of the strains being described as 'true-breeding' and 'independently-derived'? (5%)

According to her data, in the first summer Aunt Tarragon crossed each of the three strains to each other. The colours of the resulting F1 flowers are shown in Table 2.4.1.

Table 2.4.1 Flower colours of true-breeding parents and their F1 offspring.

Parents	F1
blue × red	all purple
yellow × red	all orange
blue × yellow	all blue

Aunt Tarragon is a little confused by the data, so she looked at the lily pigment database on LilyNet. The known lily pigment pathways are reproduced in Fig. 2.4.3. However, Aunt Tarragon is none the wiser and asks for your help.

'RED' pathway

White   ⟶   Red

'BLUE' pathway

White   ⟶   Yellow   ⟶   Blue

Figure 2.4.3   Known pigment synthetic pathways in the lily. Each arrow represents a single gene encoding a single enzyme. For example, a single gene encoding a single enzyme is responsible for the conversion of yellow pigment to blue pigment. In total, three genes, and therefore three enzymes, are involved.

Question 2: Given that you know the organisation of the biochemical pathways that generate lily flower colour, what is the simplest genetic explanation of the three F1 lily flower colours described in Table 2.4.1? (10%)

Question 3: Regarding Table 2.4.1, if we were to cross two F1 blue flowering lilies together, what would be the ratio of flower colours in their offspring? (5%)

Question 4: Regarding Table 2.4.1, if we were to cross two F1 orange flowering lilies together, what would be the ratio of flower colours in their offspring? (5%)

Question 5: Regarding Table 2.4.1, if we were to cross two F1 purple flowering lilies together, what would be the ratio of flower colours in their offspring? (15%)

Having resolved the genetic basis of lily flower colour, you now investigate whether there is a genetic basis to resistance and susceptibility to the viral infection lilyrot. Your aunt's well-documented studies are complicated by the necessity to store tissue from the infected lilies under strict conditions that conform to the appropriate heath and safety guidelines (see Fig. 2.4.4). However, as a result of a complex series of crosses your aunt has undertaken, it soon becomes apparent that the lilies carrying the serrated leaf mutation *savage* were nearly always resistant to lilyrot. Fortunately the DNA sequence of the *savage* gene has already been identified and maps to a well-characterised region of chromosome 12.

Figure 2.4.4   Material from lilies infected with lilyrot are only stored under very strict conditions.

Aunt Tarragon initiated a crossing scheme using true-breeding strains. She crossed a lilyrot-sensitive but otherwise wild-type strain, to a lilyrot-resistant *savage*-mutant strain and collected the F1 plants, all of which had wild-type leaves and were lilyrot-sensitive. The following season, she undertook a backcross, mating the F1 (wild-type leaf, lilyrot-sensitive) plants to homozygous recessive (*savage*-leaf, lilyrot-resistant) plants. The results of the backcross are shown in Table 2.4.2.

Question 6: Using appropriate notation, draw the full crossing scheme for the backcross undertaken by your aunt. (5%)

Table 2.4.2   Number of offspring from a series of backcrosses between F1 (wild-type leaf, lilyrot-sensitive) plants and homozygous recessive (*savage*-leaf, lilyrot-resistant) plants.

		Leaf phenotype	
		Wild-type	*savage*
Lilyrot phenotype	**Sensitive**	320	210
	Resistant	190	280

$$\chi^2 = \sum \frac{(O-E)^2}{E}$$

Figure 2.4.5   The chi-squared formula.

You decide to test Aunt Tarragon's theory that the leaf shape gene (wild-type vs *savage*) and *lilyrot* gene (sensitive vs resistant) are on the same or different chromosomes, and correctly decide to undertake a chi-squared test (see Fig. 2.4.5) on the backcross data to investigate this.

Question 7: Use a chi-squared test to determine whether the leaf-shape and *lilyrot* genes are on the same or different chromosomes. (10%). Note that for one degree of freedom chi-squared scores above 3.84 are considered significant ($p < 0.05$).

Question 8: Calculate the map distance between the leaf-shape and *lilyrot* genes. (5%)

Much to your Aunt's delight, you are able to map the *lilyrot* resistance mutation to a fairly precise chromosomal region.

Question 9: Describe a series of experiments that would allow you to identify the DNA sequence of the mutant gene which determines lilyrot resistance. How could you prove that this was the gene that, when mutant, caused the lilyrot-resistant phenotype? (10%)

You identify a cell-surface, membrane-bound receptor gene as the cause of lilyrot resistance. The lilyrot resistance mutation has a four base-pair deletion within the coding region.

Question 10: How might a four base-pair deletion in the coding region of a cell-surface, membrane-bound receptor gene lead to lilyrot resistance? How might you test your theory? (10%)

You sequence the lilyrot cell-surface, membrane-bound receptor gene in a variety of lily populations in Aunt Tarragon's Garden Centre. You find four novel mutations in the lilyrot gene. These are listed in Table 2.4.3.

Table 2.4.3   The molecular basis of four new mutations in the lilyrot cell-surface membrane-bound receptor gene.

Designation	Type of mutation
mutant 2	nonsense mutation
mutant 3	A to G change in the promoter region
mutant 4	missense mutation
mutant 5	3 base-pair insertion into the coding region

Question 11: For each of the four mutations listed in Table 2.4.3, speculate whether they might cause a lilyrot-resistant phenotype. In addition, speculate whether the mutations are likely to be dominant or recessive when heterozygous with the wild-type (lilyrot-sensitive) allele. (10%)

Question 12: Now that you understand the genetic basis of colour, leaf-shape and lilyrot resistance, how would you generate lilyrot-resistant plants (a) by creating genetically modified, transgenic lilies and (b) without creating genetically modified, transgenic lilies? (10%)

Chapter 2.5

Strawberry Fields Forever

It wasn't the summer job you'd been hoping for. After two years studying genetics at university you'd hoped to get a summer placement in a prestigious genetic research laboratory. However, despite writing an endless series of emails telling potential supervisors how entirely wonderful you are, you've had absolutely no success. Your mood does not improve when your mother tells you she's found you a summer job at Reginald Crundall's Garden Centre, as shown in Fig. 2.5.1. You are not impressed as you've worked for Reg for the last four summers and were hoping for something a little more challenging. You approach your first day with little genuine enthusiasm.

Figure 2.5.1 An impressive array of outdoor plants at Reginald Crundall's Garden Centre.

Genetics? No Problem!, First Edition. Kevin O'Dell.
© 2017 John Wiley & Sons Ltd. Published 2017 by John Wiley & Sons Ltd.

The eponymous Reg is a very experienced plant breeder and is particularly proud of the absolutely delicious strawberry, *Fragaria esculentus* var. *crundalli*, which he generated by crossing several well-established domesticated strawberry varieties. However, Reg's new strain of strawberry is quite slow growing and is easily outcompeted by other, faster-growing strawberry plants. In addition, as shown in Fig. 2.5.2, *Fragaria esculentus* var. *crundalli* have yellow fruit rather than the traditional red, which in itself is apparently less attractive to potential customers. Therefore Reg can't sell *Fragaria esculentus* var. *crundalli* as a commercial crop.

Figure 2.5.2   Traditional red-colour variants of strawberries from Reginald Crundall's Garden Centre.

Knowing that you have a good understanding of genetics, Reg lets you into a little secret. Two years ago he set up a research project with one of the world's leading plant molecular biology laboratories (see Fig. 2.5.3). The primary investigator, Professor Marie Nation, is an expert in the genetics of crop improvement, particularly with respect to taste, and she agreed to help Reg modify *Fragaria esculentus* var. *crundalli* to make it into a viable and valuable crop plant.

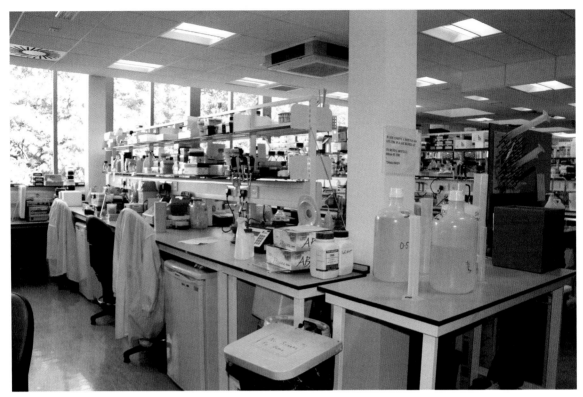

Figure 2.5.3 Professor Marie Nation's well-resourced Transgenic Plant Laboratory.

Fortunately the collaboration has recently born fruit and Professor Nation has just sent Reg some seeds from six new, genetically modified strains of *Fragaria esculentus* var. *crundalli* that she claims should solve Reg's problems. Reg has grown some of these new seeds in a controlled and restricted environment in the cellar of the garden centre and is now rather nervous about precisely what he will discover. Professor Nation has sent Reg lots of documentation about the six genetically modified strains (which she has called Reg-A to Reg-F), but Reg's degree in landscape gardening hasn't really prepared him for this. So he asks you to help.

According to the documentation, Professor Nation has made genetically modified strawberries using the standard *Agrobacterium* Ti plasmid strategy that generates random transgene insertions. In this case, she has cleverly designed a single Ti plasmid construct that, when inserted into the diploid *Fragaria esculentus* var. *crundalli* germ line, will address both the issue of slow growth and the problem with fruit colour.

To generate the transgenic strains, floral tissues from *Fragaria esculentus* var. *crundalli* were dipped into a solution containing *Agrobacterium* and the T-DNA was transferred into the strawberry germ line. The process of strawberry transformation is shown in Fig. 2.5.4.

Figure 2.5.4 The process by which *Agrobacterium* T-DNA can be used to introduce foreign DNA into the germ line of the *Fragaria esculentus* var. *crundalli*. On average this process results in approximately five insertions per plant genome.

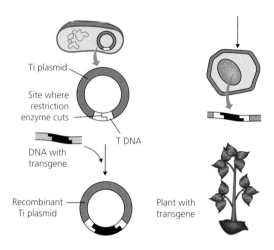

Question 1: The DNA that is transferred from the recombinant Ti plasmid to the strawberry germ line has been designed to address the issues of slow growth and fruit colour. Draw a cartoon of the piece of DNA that is transferred to the host plant (coloured grey and black in Fig. 2.5.4), clearly stating the function of each section of the DNA being transferred. (10%)

Question 2: Name two distinctly different ways in which you could determine that the new DNA had arrived in its new strawberry host? (10%)

Question 3: How would you identify the DNA sequence flanking the insertion site of the modified T-DNA in its new strawberry host? (5%)

You quickly and excitedly undertake preliminary studies of the six new transgenic plants and notice that whilst the strawberries in five of the transgenic strains (Reg-A, Reg-B, Reg-D, Reg E and Reg-F) have a deep-red colour like normal strawberries, the strawberries of strain Reg-C have more of an orange colour. Reg asks you why that might be, because Professor Nation has clearly shown that all six transgenic strains carry the correct transgenes in their genome and sequence analysis clearly shows that neither of the transgenes have mutated.

Question 4: Suggest two possible explanations for strain Reg-C having orange fruit. How would you distinguish between these explanations? (10%)

You change your focus to investigate the growth rate of the six new transgenic plants and are delighted that they all seem to grow more quickly than the original *Fragaria esculentus* var. *crundalli* strain. Detailed analyses of the growth rates are shown in Fig. 2.5.5.

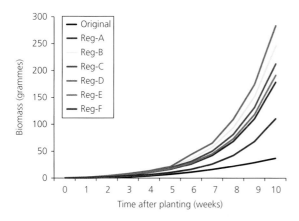

Figure 2.5.5 Increase in average biomass of the original yellow fruiting, slow growing but tasty strawberries (*Fragaria esculentus* var. *crundalli*) from Reg's Garden Centre for the first ten weeks after planting, relative to the six transgenic strains constructed by Professor Nation.

You show your growth rate data to Reg and whilst he is delighted that all six transgenic strains grow significantly faster than the original *Fragaria esculentus* var. *crundalli* strain, he is surprised that their grow rates are so different. After all, each of them have the same transgene in the same original host.

Question 5: Your first thought is that the different transgenic strains may have different transgene copy numbers. How could you investigate this theory? (5%)

In fact, you discover that all the transgenic lines (Reg-A to Reg-F) have a single transgene insertion.

Question 6: Speculate on how a single insertion of the same transgene at different positions in identical host genomes might result in the different growth rate phenotypes seen in Fig. 2.5.5. (10%)

Reg is delighted, especially as all six transgenic strains of *Fragaria esculentus* var. *crundalli* seem to have retained their original delicious taste. You organise a taste test at the garden centre and his customers overwhelmingly vote for transgenic strain Reg-B as their favourite.

Question 7: Briefly describe how you would ensure that that the taste test produces reliable results? (5%)

However, there is a problem with strain Reg-B, in that their growth rates are extraordinarily variable, whereas all the other transgenic strains exhibit fairly consistent growth rates. In fact, as far as Reg-B is concerned, Fig. 2.5.5 is quite misleading. Upon closer inspection you discover there seem to be two distinct types of Reg-B plants, those that grow slowly and are indistinguishable from the original *Fragaria esculentus* var. *crundalli* (which you call Reg-B slow) and those that are in one of the fastest growing transgenic strains (Reg-B fast). You cross these Reg-B variants to each other to investigate the genetic basis of these two distinct phenotypes. These data are shown in Table 2.5.1.

Table 2.5.1   The F1 offspring resulting from crosses between the two different classes of Reg-B strains.

Parents	F1 offspring
slow × slow	all slow
fast × slow	50% fast: 50% slow
fast × fast	67% fast: 33% slow

Question 8: Speculate of why there are two different classes of Reg-B strain. (10%)

Question 9: How might you generate a true-breeding, fast-growing Reg-B strain? (10%)

Whilst the plant growing facilities at Reg's garden centre are magnificent, you are surprised to learn that he doesn't possess a state-of-the-art molecular genetics laboratory, so you contact one of your university lecturers, Professor Jacqueline Hyde, who you know is working on the *giant* gene of *Fragaria* species. Professor Hyde has shown that the *giant* gene is a key regulator of growth rate in strawberries. As this is a very recent discovery and has only just been published, you know for certain that the *giant* gene is not involved in the genetic modification of the six transgenic *Fragaria esculentus* var. *crundalli* constructed by Professor Nation.

Professor Hyde kindly agrees to use qPCR to investigate expression levels of *giant* in the original *Fragaria esculentus* var. *crundalli* strain as well as the six transgenic strawberries (Reg-A to Reg-F), including the two versions of Reg-B (fast and slow) described earlier. Professor Hyde's results are shown in Fig. 2.5.6.

Figure 2.5.6 qPCR analysis of the *giant* gene from the original *Fragaria esculentus* var. *crundalli* strain and the six transgenic strawberries (Reg-A to Reg-F) including the two versions of Reg-B (fast and slow) described earlier.

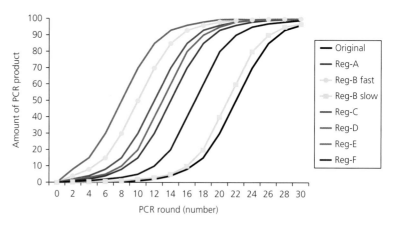

Question 10: How does qPCR enable you to investigate gene expression levels? (5%)

Question 11: What is the relationship between *giant* expression and plant growth rate? (5%)

Question 12: Speculate on the mechanism by which *giant* expression is affected by the presence of the transgene introduced by Professor Marie Nation. How would you test this? (15%)

Chapter 2.6

The Mystery of Trypton Fell

You have just completed your first year at university and have decided your future lies in a career as a forensic scientist. That is why you're already studying genetics at university, as it's just the kind of qualification you'll need. However, you are acutely aware that to reach your goal of being a forensic scientist you're going to need more than just an excellent degree at an excellent university. You need some relevant work experience.

Over the past few months you've been applying for summer jobs at a variety of forensics laboratories, unfortunately without success. It would seem that they are all overwhelmed with students wanting summer placements and most of the competition is further on in their degree studies than you. Nevertheless, you are delighted to receive a text from the partners of the revered law firm *Interpret, Represent and Evidence* who are keen for you to be their slave

Figure 2.6.1   Trypton Fell, stately home of Lord and Lady Trypton.

Genetics? No Problem!, First Edition. Kevin O'Dell.
© 2017 John Wiley & Sons Ltd. Published 2017 by John Wiley & Sons Ltd.

during the summer break. You hope that at some point you'll be given a chance to demonstrate your talent.

On your first day you are stunned by the opulent surroundings and begin to realise that they must be one of the most successful law firms in the country. For much of the first week, as you rather expected, you make lots of cups of tea, which at least gives you the chance to meet the three senior female partners, Miss Interpret, Miss Represent and Doctor Evidence. On that first Friday there is a real excitement in the air as one of their most prestigious clients, Lord Trypton, is rumoured to have entered the building.

You remember reading all about Lord Trypton on Aristonet, the website dedicated to the few surviving members of the British aristocracy. He owns the vast estate of Trypton Fell, which is shown in Fig. 2.6.1. His family fortune is based on developing rare specialist breeds of farm animals, especially goats, pigs and sheep. However, today he is also a very angry man. 'It's the Bottomless-Pitts,' he screams. 'My neighbours. They've gone too far this time and killed old Porky!'

You immediately realise this may be a great opportunity to impress future employers by demonstrating your potential as a forensic geneticist and you set about solving the mystery of Trypton Fell.

Later that afternoon you are given the opportunity to visit the scene of the alleged crime. The body of Porky still lies where he was originally found. Whilst you're surveying the scene looking for clues, a stylish young woman, very elegantly dressed and wearing green Wellington boots, comes over to talk to you. She is clearly quite distraught. 'I'm Lady Mary,' she says. 'Welcome to Trypton Fell. I hear you're the forensics expert. I really hope you solve our problem. So many of our pigs have died recently.'

You have a long conversation with Lady Mary and you're impressed by her knowledge of all thing piggy. She explains that a lot of apparently healthy pigs have died suddenly and she hands you her meticulous notes on the subject. You look at the data and soon realise that though Porky is now dead, while he was alive he had a particularly active life. A typical pedigree analysis showing some of Porky's offspring is shown in Fig. 2.6.2.

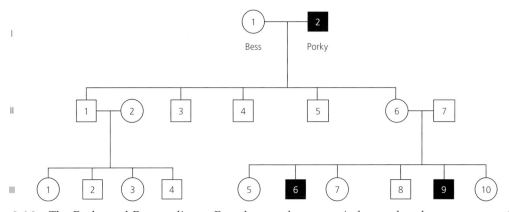

Figure 2.6.2   The Porky and Bess pedigree. Females are shown as circles, and males are squares. Dark squares and circles denote pigs that died suddenly and unexpectedly.

Several pedigrees you look at are all broadly similar as nearly all the pigs that died suddenly were male. You begin to think this isn't a crime perpetrated by Darren Bottomless-Pitt and may actually be associated with an X-linked genetic condition.

Question 1: What is the evidence that the pigs are dying from an X-linked genetic condition? (10%)

Question 2: Lady Mary tells you that female II:6 suffered from occasional unexplained paralysis, then died at a respectable old age after an unfortunate incident with a runaway tractor. Interestingly, one of her daughters (III:7) also exhibited occasional unexplained paralysis. Why might female II:6 and her daughter III:7 have suffered from occasional unexplained paralysis? (10%)

You study Lady Mary's data in more detail. You notice that male pigs are either black or pink, whereas females can be black, pink or 'blotchy'. Blotchy females have random patterning with large areas of black and pink. A typical pink female is in shown in Fig. 2.6.3.

Figure 2.6.3 A sow from the Trypton Fell Estate exhibiting the pink body colour and normal ear phenotypes.

Question 3: Speculate on the genetic basis of coat colour in Lady Mary's pigs. (10%)

You decide to contact Professor Susie Trotter, the world porcine expert. Professor Trotter is fascinated by your work and is sad to hear of Porky's death. She had suspected Porky might be susceptible to sudden unexplained death and has been investigating the genetic basis of the condition. She believes that susceptible pigs only die when they are very hungry. She has been trying to map the sudden unexplained death gene to a precise position on the X-chromosome using a three-point test cross. This involves finding the position of the sudden explained death gene relative to the body colour gene described earlier and a gene that affects ear shape where the normal ear shape allele was dominant to the ragged ear allele.

Professor Trotter mated Porky (who was black and had normal ears) to Peppa (who was pink with ragged ears and had no family history of sudden unexplained death). They produced three sons and six daughters. She then took the 6 daughters (all of whom had blotchy colouring and normal ears) and noticed no sudden deaths when the pigs were deprived of food for 24 hours. She then mated the six daughters to random males multiple times. She collected data on the phenotypes of the male offspring, which is shown in Table 2.6.1.

Question 4: Using appropriate genetic notation, draw Professor Trotter's crossing scheme, showing the genotype of each individual. (10%)

Question 5: It is stated that the F1 triply heterozygous mother can be mated to any random male. Why doesn't the genotype of the males (the fathers) matter and why do you only use their sons to calculate the map distance? (10%)

Table 2.6.1   Offspring from a series of testcrosses between triply heterozygous F1 mothers and random males as described in the text.

Sudden death gene	Colour gene	Ear shape gene	F2 males
susceptible	black	normal	205
resistant	pink	ragged	193
susceptible	pink	normal	17
resistant	black	ragged	13
susceptible	pink	ragged	1
resistant	black	normal	2
susceptible	black	ragged	36
resistant	pink	normal	33

Question 6: Calculate the map distances between the three X-linked genes and draw a map to show their relative positions. (10%)

Professor Trotter is delighted that you'd found the approximate map position of the sudden unexplained death gene but there are still a lot of genes within that X-chromosome region.

Question 7: How could you find the DNA sequence of the mutant sudden unexplained death gene within the pig X-chromosome? What problems might you encounter? (10%)

You return to the scene of the crime at Trypton Fell to discuss your progress on the case with Lord Trypton and Lady Mary. His Lordship, who is still very angry, doesn't believe your genetics theory, but Lady Mary is very interested indeed.

Suddenly, out of the corner of your eye you notice a whole litter of piglets in a state of distress. You rush over to help and remember Professor Trotter saying that the sudden unexplained death phenotype is associated with starvation. She advises feeding the pigs as soon as possible and then keeping them warm for 24 hours to help them survive.

The only food you have with you are some bananas, and with the help of Lady Mary you force some banana down the throat of the distressed young animals. To everyone's astonishment the piglets recover within minutes and, as shown in Fig. 2.6.4, after a good night's sleep they seem to be perfectly healthy and none the worse for their ordeal.

Figure 2.6.4   The sleeping piglets after their near-death experience.

You remember receiving a lecture about this, where horses with a sodium channel mutation would recover after being fed potassium-rich foods such as bananas. In that case, the horses had a mutation in one of their sodium channel genes and you wonder whether the sudden unexplained death phenotype in the pigs is associated with a sodium channel mutation. Interestingly, the porcine genome project reveals that there are three pig sodium channel genes in the region of the X-chromosome you've already identified as causing the sudden unexplained death phenotype.

Question 8: How are you going to prove which, if any, of the sodium channel genes is the cause of the mutant phenotype? What problems might you encounter? (15%)

The three lawyers are hugely impressed by your work and delighted that you've solved the mystery of Trypton Fell. Lady Mary is impressed too and although she assures you her father is similarly impressed, he still seems to be very angry.

Why is there so much animosity between Lord Trypton and Darren Bottomless-Pitt? Surely it can't just be because he is disappointed that his daughter Sybil seems to have a bit of a crush on Darren's son, Keanu Bottomless-Pitt, who is employed as a chauffeur to his Lordship? In fact you have suspected all along that Trypton Fell has another secret, so at the same time you've been investigating the fate of poor Porky, you've also been surreptitiously collecting DNA samples from each member of the family. You return to Professor Trotter's laboratory and determine the full 13-site STR profiles of the Tryptons and the Bottomless-Pitts. These data are presented in Table 2.6.2.

Table 2.6.2 DNA Profile of every surviving member of the Trypton and Bottomless-Pitt families, using 13 pairs of STRs (numbered STR1 to STR13). The numbers in the table represent the STR profile at each specific STR. For example at STR1 Dowager Countess Trypton has a STR profile of 21 and 26 repeats.

Character	Age	STR1	STR2	STR3	STR4	STR5	STR6	STR7	STR8	STR9	STR10	STR11	STR12	STR13
Dowager Countess Trypton	72	21/26	18/23	42/46	33/34	52/57	22/26	44/44	23/30	67/68	82/87	17/19	56/61	14/14
Lord Trypton	58	21/22	20/23	42/43	33/36	55/57	22/27	44/50	26/30	64/67	85/87	19/19	56/60	14/19
Lady Trypton	53	24/25	19/24	43/48	35/38	53/54	23/24	46/49	29/30	66/66	81/86	14/15	58/62	14/14
Lady Mary Trypton	26	21/25	20/24	43/43	33/35	53/57	24/27	49/50	26/30	64/66	81/87	14/19	56/58	14/14
Lady Edith Trypton	24	22/25	19/23	43/48	36/38	53/55	23/27	44/49	30/30	64/66	86/87	14/19	58/60	14/14
Lady Sybil Trypton	22	22/24	19/20	42/48	36/38	53/57	22/23	44/46	26/29	66/67	86/87	15/19	60/62	14/14
Darren Bottomless-Pitt	39	20/23	21/22	41/44	32/35	58/59	27/30	47/51	27/28	61/63	80/81	18/20	59/63	14/19
Sharon Bottomless-Pitt	38	20/26	16/21	45/47	32/40	51/57	25/29	51/51	24/25	61/64	79/83	13/18	57/62	14/14
Keanu Bottomless-Pitt	21	20/21	21/23	42/47	33/40	55/57	22/25	50/51	25/30	64/67	83/85	18/19	60/62	14/19
Chardonnay Bottomless-Pitt	16	23/26	21/21	44/47	32/32	57/59	29/30	51/51	24/27	63/64	81/83	13/18	62/63	14/14

Question 9: Who are the parents of: (a) Lady Mary; (b) Lady Edith; (c) Lady Sybil; (d) Keanu Bottomless-Pitt; (e) Chardonnay Bottomless-Pitt. (5%)

Question 10: Darren and Sharon Bottomless-Pitt are not actually married. Explain why they may share the same family name? (5%)

Question 11: Why might Lord Trypton be particularly concerned at the relationship between Lady Sybil and Keanu Bottomless-Pitt? (5%)

Sir Henry's Enormous Chest

It's been a bit of a disaster and it's entirely your own fault. Your lecturers had encouraged you to apply for summer laboratory placements but for some reason you never quite managed to get round to doing it. Now all your friends on the genetics degree course have managed to get placements and you're left working in a glorified junk shop, or Sir Henry's Interesting Treasure Emporium to give it its proper name.

You soon realise that by the end of the summer you'll be an expert at moving pieces of paper from one pile to another. Sir Henry's business seems to involve clearing items, especially old documents, from abandoned houses in the vain hope that something valuable turns up. Clearly Sir Henry is a bit of an expert at this because he's managed to make a highly successful career out of it, but how can you tell whether anything is valuable or not?

In your second week, you're looking through the drawers of the antique mahogany chest shown in Fig. 2.7.1, when you discover a pile of old medical documents, one of which seems to resemble a family pedigree. On closer inspection, you notice that next to the names are a single letter code, namely A, B or O. You also notice that several of the squares (which seem to be males) and circles (which seem to be females) have been coloured in. It's only when you see the inscription 'Renwick, NPS 1956' in the top right-hand corner that you realise you are holding in your hand an original pedigree from James Renwick's laboratory.

Figure 2.7.1 Scene from inside Sir Henry's Interesting Treasure Emporium, showing the antique chest of draws within which you discovered a pile of old documents.

Genetics? No Problem!, First Edition. Kevin O'Dell.
© 2017 John Wiley & Sons Ltd. Published 2017 by John Wiley & Sons Ltd.

But who was James Renwick? You remember being told that he worked at the University of Glasgow in the 1950s and was one of the first people to find the chromosomal position of a disease-causing mutation. You gently put the pedigree back in its envelope, place it carefully inside your pocket, tell Sir Henry that you've suddenly been taken terribly ill and make your way home.

You spend most of the afternoon online researching the Renwick pedigree that you've found. 'NPS' must stand for nail–patella syndrome, a very rare condition characterised by abnormalities in nails and patellae (knee-caps), as shown in Fig. 2.7.2.

Figure 2.7.2 Typical nail (left) and patella (right) phenotypes of an individual affected by nail–patella syndrome (NPS).

The following day, you remember that as part of your degree assessment you need to write a dissertation during the summer vacation. You decide that this original NPS pedigree data of Renwick's is a potentially fascinating start point and begin planning what you hope will be a highly original dissertation. Calling it *NPS: Past, Present and Future* seems a good place to start. The pedigree that you found is reproduced, along with the ABO blood group and NPS phenotypes, in Fig. 2.7.3.

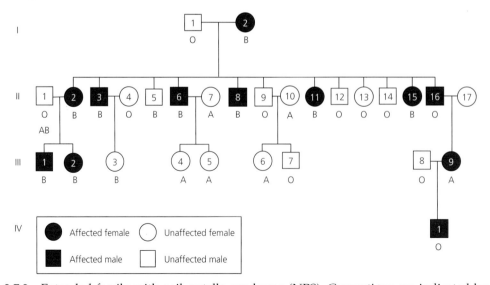

Figure 2.7.3 Extended family with nail–patella syndrome (NPS). Generations are indicated by roman numerals (I, II, III and IV), whilst individuals within that generation are numbered 1, 2 etc.

You start your dissertation by looking at the data exactly as you imagine Renwick might have done in the late 1950s and prepare several questions based on the pedigree you found.

Question 1: What is the evidence that NPS is an autosomal dominant condition? (5%)

Question 2: Redraw Fig. 2.7.3 showing, wherever possible, the ABO genotypes for all members of the pedigree. (10%)

Question 3: Are there any individuals whose ABO genotype cannot be established? If so, who are they, why can't their ABO genotype be determined, and how could you resolve this? (10%)

Question 4: What is the evidence that NPS is genetically linked to ABO? (5%)

Question 5: Calculate the map distance between NPS and ABO. (10%)

Question 6: If male III:1 from the pedigree and an unaffected woman with blood group O have a child, what is the probability that it will be a boy, have blood group B and have nail–patella syndrome? (5%)

Question 7: If we found another pedigree apparently showing inheritance of a condition with an NPS-like phenotype, would we be justified in combining the data from the two families? What might make you hesitate over such a decision? (5%)

You're pleased with the progress of the first half of your dissertation and you decide that the second half should put Renwick's data in a modern, molecular genetic context. Over the next few days you read a lot of contemporary literature on NPS and, rather like Renwick's pedigree data, it raises a number of questions that you are keen to address.

The NPS gene, which the human genome project now refers to as LMX1B, has been found to encode a transcription factor that plays an important, but not very well understood, role in the development of limbs, nails, kidneys and eyes. DNA sequencing of descendants of the family in Fig. 2.7.3 has revealed that the dominant NPS condition identified by Renwick is caused by a single base pair change.

Question 8: Speculate on how a single base pair change in the coding region of the NPS transcription factor gene could result in a dominant mutation leading to the NPS mutant phenotypes. (5%)

Question 9: Speculate on how a single base pair change in a non-coding region of the NPS transcription factor gene could result in a dominant mutation leading to the NPS mutant phenotypes. (5%)

Although NPS is described as an autosomal dominant condition, there is considerable variation in the severity of the condition, even when an identical NPS mutation is segregating within a specific family. Indeed, NPS can appear to 'skip' entire generations within a family.

Question 10: Speculate on why the same mutation in different members of the same family causes a different severity of phenotype in different individuals. (10%)

The literature also reveals that standard diagnostic tests designed to identify mutations in the NPS gene is successful in about 85% of affected families.

Question 11: Why might standard diagnostics tests only identify 85% of affected families? (5%)

You wonder whether it might be possible to develop a treatment or cure for NPS by modelling the disorder in mice. You wonder what would happen if you were to take the dominant mutant version of the human NPS gene and introduce it to a host wild-type mouse. You contact researchers in a laboratory that have attempted this. They can prove that the human NPS transgene has successfully integrated, that its sequence has not been modified in any way, and that it is expressed in the correct cell types. Nevertheless, these transgenic mice, shown in Fig. 2.7.4, have absolutely no mutant phenotype.

Figure 2.7.4 A wild-type mouse carrying the human NPS mutant transgene that is apparently unaffected by NPS.

Question 12: Speculate on why wild-type mice carrying the human NPS mutant transgene do not have a mutant phenotype. (5%)

Other laboratories have used an alternative transgenic approach. They start with the homologous mouse NPS gene, prior to introducing it to host wild-type mice.

Question 13: Prior to introducing it to a wild-type mouse embryo, what modifications should they make to the mouse NPS gene that would enable them to investigate why the original NPS mutation identified by Renwick is dominant. What problems might they encounter? (10%)

This time the transgenic mice show a clear NPS phenotype.

Question 14: Given that you now have a transgenic mouse model, what experimental strategy would you use to find a treatment for human NPS? (10%)

Chapter 2.8

Pandemonium

Your friends are insanely jealous but you have absolutely no sympathy with them. After all, any of them could have applied for a summer project at the Chengdu Research Base of Giant Panda Breeding in China (see Fig. 2.8.1) but only you had the initiative to actually do so. Now you're flying into Chengdu Shuangliu International Airport for an all-expenses-paid summer research placement in the world famous laboratory of Professor Feng Shui.

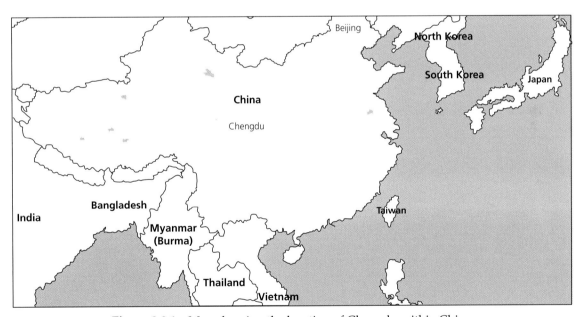

Figure 2.8.1 Map showing the location of Chengdu within China.

One of the main reasons Professor Shui picked you from the hundreds of applicants was because of your dissertation on *Captive Chimpanzee Breeding Programmes* that you've undertaken as part of your studies for your genetics degree (see Fig. 2.8.2). She believes that you have an

Genetics? No Problem!, First Edition. Kevin O'Dell.
© 2017 John Wiley & Sons Ltd. Published 2017 by John Wiley & Sons Ltd.

excellent understanding of how breeding programmes should be organised and an appreciation of the problems that may be encountered. You decide not to tell her that your dissertation was very much a team effort and was only possible because so many of the researchers you collaborated with were pulling in one direction.

Figure 2.8.2   Chimpanzees at the captive breeding programme.

On the flight to Chengdu you spend most of the time reading over your award-winning dissertation on *Captive Chimpanzee Breeding Programmes*. One key aspect of the programme is establishing parentage within captive chimpanzee troops. A female chimpanzee may mate with several males so it important to use forensic DNA-typing techniques to establish true genetic parentage. Researchers use short-tandem repeats (STRs) to establish chimpanzee parentage beyond all reasonable doubt.

As it says in your dissertation, STRs are short sequences of DNA that are usually three or four base pairs long. They are found at specific points on a chromosome and are repeated several times. In your dissertation, for example, chimpanzee STR A is a CGT trinucleotide repeat on chromosome 2 that varies from 5 to 15 tandem repeats. Because each chimpanzee has two chromosome 2's, they have two STR A's. Chimpanzees with STR A genotype 5/7 have five repeats of CGT on one chromosome 2 (CGTCGTCGTCGTCGT) and seven on their other chromosome 2 (CGTCGTCGTCGTCGTCGTCGT).

You studied a captive female chimpanzee called Jane who had just given birth to an infant. Prior to conception she had the opportunity to mate with at least five different and unrelated captive males and you were keen to determine which of them had the 'X-factor'. You decided to investigate variation at 10 STRs (STRs A–J) to establish paternity and these data are shown in Table 2.8.1.

Table 2.8.1 STR profiles of seven chimpanzees at 10 unlinked STRs. Jane is the mother of the infant chimpanzee. Harry, Liam, Louis, Niall and Zayn are possible fathers of the infant.

STR	Jane	Harry	Liam	Louis	Niall	Zayn	Infant
STR A	7/10	8/9	12/15	11/13	9/10	7/11	7/10
STR B	11/13	12/17	14/15	12/15	12/16	10/13	11/16
STR C	26/31	24/24	29/30	28/31	25/27	26/30	25/31
STR D	21/23	19/23	24/26	20/25	22/25	25/26	22/23
STR E	24/25	28/31	24/29	27/29	25/30	26/29	25/30
STR F	16/17	12/17	14/14	15/18	13/14	12/18	14/16
STR G	23/23	19/22	20/24	21/22	21/23	19/24	21/23
STR H	8/14	11/11	9/12	10/13	12/13	8/9	12/14
STR I	32/33	31/38	34/35	36/36	34/37	35/39	33/37
STR J	8/12	8/9	10/13	9/10	11/14	12/15	11/12

Question 1: Who is the father of the infant? What is the evidence for this? (10%)

You are excited by the prospect of working with giant pandas and are pleased that the Chengdu researchers also use STRs to help determine parentage. In Chengdu they use 13 STRs and the researchers in Professor Shui's laboratory call them STR 1–13.

On arriving in Chengdu you have your first opportunity to discuss your research project with Professor Shui. One panda she is particularly excited about is Yin Yang, a wild-caught giant panda from Shaanxi province. Yin Yang is of special interest as the giant pandas from Shaanxi province are quite rare and the researchers are keen to introduce them into their captive breeding programme, which is currently very dependent on giant panda's from Chengdu.

Question 2: Why might giant pandas from a different Chinese province be a valuable addition to the Chengdu breeding programme? (10%)

To ensure that all their pedigree analyses are as complete as possible, Professor Shui and her colleagues have collected a lot of observational data from wild giant pandas from Shaanxi province, and have also collected bucket-loads of giant panda poo, which has allowed them to establish DNA profiles for three female and three male wild giant pandas. They are fairly sure that two of these giant pandas must be Yin Yang's parents and are keen to identify them. The STR profiles of Yin Yang, the three adult females and three adult males are shown in Table 2.8.2.

Table 2.8.2 STR profiles of three wild female and three wild male giant pandas, any of whom may be the parents of wild-caught giant panda cub Yin Yang.

Name	Female 1	Female 2	Female 3	Male 1	Male 2	Male 3	Yin Yang
STR 1	6/7	6/6	6/8	7/7	7/8	6/6	6/7
STR 2	11/11	12/13	12/13	11/12	12/12	11/13	11/13
STR 3	22/23	23/25	23/24	23/23	23/24	22/24	23/25
STR 4	10/11	9/10	11/11	11/12	10/11	9/11	9/11
STR 5	7/9	8/9	7/7	7/8	8/9	8/9	8/8
STR 6	15/17	15/17	14/16	16/16	15/15	14/15	15/16
STR 7	19/21	22/22	21/22	19/21	21/21	20/21	21/22
STR 8	15/18	15/16	15/15	15/18	17/18	15/17	15/15
STR 9	9/10	8/10	9/10	8/10	10/10	9/10	10/10
STR 10	17/17	19/20	20/20	17/19	18/20	18/20	19/20
STR 11	5/5	5/6	5/7	6/6	6/6	6/7	5/6
STR 12	17/17	17/17	17/17	17/22	17/22	17/22	17/22
STR 13	12/12	12/12	14/14	11/12	11/14	13/14	12/13

Question 3: Taking each female in turn (and at this stage ignoring the males), discuss which female(s) may be Yin Yang's mother? (5%)

Question 4: Taking each male in turn (and at this stage ignoring the females), discuss which male(s) could be Yin Yang's father? (5%)

Question 5: Which pair, if any, might be Yin Yang's parents? What issues, if any, do the results from the STR profiling raise? (10%)

Question 6: Speculate on whether any of the STRs could be used as a sex test? What sex does this suggest Yin Yang is? (10%)

During your third week in Chengdu, Professor Shui introduces you to the artificial insemination team. They explain why it's so difficult to breed captive giant pandas (see Fig. 2.8.3). Apparently females are only fertile or on-heat for two or three spring days each year and to the untrained, or even highly trained, human eye it's almost impossible to work out precisely when this is. In addition, populations of giant pandas are so inbred that some researchers believe something approaching 70% of female giant pandas are unable to bear cubs and perhaps as many as 80% of male giant pandas are sterile.

Figure 2.8.3 Captive Giant Panda awaiting artificial insemination.

Professor Shui and her colleagues are using a variety of strategies to try to increase the success rate of artificial insemination. She shows you some recent data from using a wild-caught female giant panda from Shaanxi, identified as female 1 in Table 2.8.2. The researchers inseminated her with sperm from the three wild Shaanxi male giant pandas (males 1–3), see Table 2.8.2, plus two captive-born males from Chengdu (males 4 and 5). The female has recently given birth to a cub and Professor Shui is keen to know who the father is. The full STR profiles of the seven giant pandas are shown in Table 2.8.3.

Table 2.8.3 STR profiles from giant panda blood samples taken from wild-caught female 1 and five male giant pandas (male 1-5) who were used as sperm donors for artificial insemination, one of whom must be the father of the new cub.

Sex	Female 1	Male 1	Male 2	Male 3	Male 4	Male 5	Cub
STR 1	6/7	7/7	7/8	6/6	7/7	7/8	6/7
STR 2	11/11	11/12	12/12	11/13	11/12	12/13	11/12
STR 3	22/23	23/23	23/24	22/24	22/24	23/24	23/24
STR 4	10/11	11/12	10/11	9/11	10/10	11/11	10/11
STR 5	7/9	7/8	8/9	8/9	7/8	8/8	8/9
STR 6	15/17	16/16	15/15	14/15	15/17	15/16	15/17
STR 7	19/21	19/21	21/21	20/21	19/20	20/21	19/21
STR 8	15/18	15/18	17/18	15/17	17/18	16/18	18/18
STR 9	9/10	8/10	10/10	9/10	9/10	8/9	8/10
STR 10	17/17	17/19	18/20	18/20	18/20	19/20	17/20
STR 11	5/5	6/6	6/6	6/7	6/7	5/6	5/6
STR 12	17/17	17/22	17/22	17/22	17/22	17/22	17/22
STR 13	12/12	11/12	11/14	13/14	11/12	13/14	12/14

Question 7: Who is the father of the new cub? What is the evidence for this? (10%)

Question 8: What sex is the new cub? What is the evidence for this? (5%)

Question 9: Why is it more difficult to determine parentage in giant pandas than in chimpanzees? (10%)

You are curious as to why Yin Yang, who at 6 years old is probably old enough to breed, was not included in the artificial insemination programme. You are particularly interested in the welfare of Yin Yang because prior to arriving in Chengdu, you'd been liaising with your local zoo about the hope of convincing Professor Shui that the zoo could house a male and female giant panda. Professor Shui tells you that like many males, Yin Yang seems to be infertile, but she is happy for you to spend the rest of your time in Chengdu investigating why.

So far you've only been observing the giant pandas from a distance and from what you've learned so far, they seem to behave just like human teenagers, sleeping for 20 hours a day. In fact, in many respects you found the animatronics at Harry Potter World far more convincing. This time you're going to get up close and personal with Yin Yang and you're somewhat horrified that this means dressing in a giant panda costume and covering yourself in giant panda poo.

Under the strict supervision of the giant panda vet, you undertake a full examination of Yin Yang. Later in the laboratory you make a startling discovery. Unlike all the other giant pandas that have 42 chromosomes (20 pairs of autosomes and 1 pair of sex chromosomes), Yin Yang has 43. In fact Yin Yang has an XXY chromosomal composition.

Question 10: What experiment could you do to determine the chromosomal number of a cell and, in particular, how could you identify which chromosome is present in abnormal number? (10%)

Question 11: What name is given to the human genetic XXY condition? What phenotype do human XXY males have? (5%)

Professor Feng Shui is fascinated by your discovery and she explains that XXY individuals arise due to non-disjunction of homologous chromosomes during meiosis. She provides you with a cartoon explaining non-disjunction from her latest book (which is reproduced here in Fig. 2.8.4), which shows how non-disjunction leading to abnormal sex chromosome number could occur in meiosis I or meiosis II.

You immediately realize that Yin Yang's Y-chromosome must come from his father, but the two X-chromosomes could either both come from his mother or he could get an X-chromosome from each of his parents.

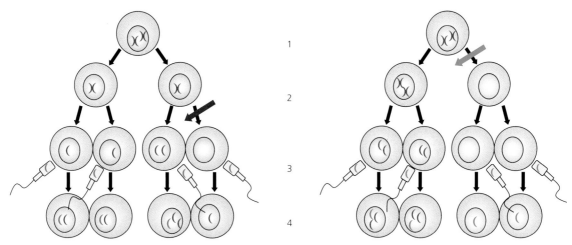

Figure 2.8.4 Nondisjunction is when chromosomes fail to separate normally resulting in a gain or loss of chromosomes. In the left image the blue arrow shows nondisjunction taking place during meiosis II. In the right image at the green arrow is nondisjunction taking place during meiosis I. 1, Meiosis I; 2, meiosis II; 3, fertilization; 4, zygote.

You undertake a more detailed sequence analysis of giant panda sex chromosomes and discover a new hypervariable STR that is found on the X-chromosome but not on the Y-chromosome. Therefore, normal giant panda females (XX) have two copies of this STR and normal giant panda males (XY) have one copy of this STR. An analysis of this new STR in Yin Yang and his parents is shown in Table 2.8.4. These data, coupled with that described earlier, confirm that Yin Yang has an XXY sex chromosome karyotype.

Table 2.8.4 STR profiles of Yin Yang and his parents at the newly discovered hypervariable X-chromosome STR.

Giant panda	New STR
mother	42/44
father	47
Yin Yang	42/42

Question 12: Speculate on the cause of the XXY karyotype of Yin Yang. In particular, can you determine whether non-disjunction occurred at meiosis I or at meiosis II of his mother or his father? (10%)

Chapter 2.9

My Imperfect Cousin

It is summer again and this means only one thing, your cousin is coming to stay. You couldn't really remember precisely when it all started but your cousin always visits in the summer. He is a perfectly reasonable person and you are about the same age, but that and family ties are about the only thing you have in common. He is very much a city boy and has an almost pathological distrust of the countryside. But this visit is probably going to be his last summer trip, as you are about to go to university to study genetics, whereas he is …? Actually you have no idea of his plans, but then again, he probably has no idea either.

You vividly remember that first summer when you were playing outside near your father's vegetable patch (see Fig. 2.9.1). 'What are those?' asked your cousin. 'Tomatoes,' you replied somewhat surprised. It had not occurred to you that your cousin would never have seen a growing tomato before.

Figure 2.9.1 Your father's vegetable patch.

Genetics? No Problem!, First Edition. Kevin O'Dell.
© 2017 John Wiley & Sons Ltd. Published 2017 by John Wiley & Sons Ltd.

Strangely, this had given you and your cousin something in common and over the following few summers you'd grown several different strains of tomato from seed, meticulously taking photographs and recording the results (see Fig. 2.9.2). In recent summers you'd been a bit more adventurous and crossed some strains together. Before your cousin arrives for what will probably be his final visit, you decide to look through your notebook and check up on your results from the past few years.

Figure 2.9.2   Round (left) and oval (right) tomatoes grown in your father's vegetable patch.

In the first year that your cousin visited, your father provided you with seeds from the three true-breeding tomato strains listed in Table 2.9.1.

Table 2.9.1   The original three true-breeding tomato strains supplied by your father.

Strain	Tomato phenotype
1	red round
2	yellow oval
3	yellow oval

Along with your cousin, you crossed strains 1 and 2 together and obtained an F1 where all the tomatoes were round and red.

Question 1: Propose a theory that explains why the offspring of a cross between strain 1 (red round) and strain 2 (yellow oval) produces F1 tomatoes that are all red and round. (5%)

The following summer, the two of you crossed the F1 plants together and produced four classes of F2 tomatoes. The numbers of each class of F2 tomato are shown in Table 2.9.2.

Table 2.9.2   Number of F2 tomatoes with each phenotype.

Tomato phenotype	Number
red round	571
red oval	176
yellow round	193
yellow oval	60

Question 2: Propose a theory that explains the genetic basis of tomato colour and shape as suggested by data from the second cross that are shown in Table 2.9.2. (5%)

You correctly decide to test your theory statistically with a chi-squared test. The chi-squared formula is shown in Fig. 2.9.3. For three degrees of freedom, the 5% significance limit is 7.185.

$$\chi^2 = \Sigma \frac{(O-E)^2}{E}$$

To assist in calculating the chi-squared value to help test your theory, you may wish to redraw Table 2.9.3.

Figure 2.9.3   The chi-squared formula.

Table 2.9.3   Use the above table to calculate the chi-squared value for the theory you proposed in the answer to question 3.

Phenotype	Observed number	Expected number	$(O - E)$	$(O - E)^2$	$\dfrac{(O - E)^2}{E}$
red round	571				
red oval	176				
yellow round	193				
yellow oval	60				
TOTAL	1000				

Question 3: Use Table 2.9.3 to calculate the chi-squared value to test whether your theory is plausible. What do you conclude? (15%)

In parallel with the experiment crossing strains 1 and 2, you also crossed strains 1 and 3 and obtained data that were almost identical. You sensibly conclude that the genetic basis of tomato colour and shape is the same in each strain and that strains 2 and 3 are identical, in that they have mutations in the same shape and colour genes.

Question 4: If strains 2 and 3 have mutations in the same shape and colour genes, what would be the phenotypes of their F1 and F2 offspring? (5%)

You crossed the yellow oval tomato strains 2 and 3 together and surprisingly the resulting F1 tomatoes were all yellow and round. You kept these rather strange results a secret from your cousin and the following summer, you crossed the F1 tomatoes together and generated the plants shown in Table 2.9.4.

Table 2.9.4   Number of F2 Tomatoes with each phenotype as a result of a cross between F1 parents derived from original true-breeding yellow oval parents from strains 1 and 2 as described above.

Tomato phenotype	Number
yellow round	565
yellow oval	435

Your first thought was that this was a 1:1 ratio, though you couldn't understand why.

Question 5: Use Table 2.9.5 to help you calculate the chi-squared value and test whether the data in Table 2.9.4 are in a 1:1 ratio. In this case there is one degree of freedom and the 5% cut off is 3.84. (10%)

Table 2.9.5   Use this table to calculate the chi-squared value to test the theory that the two classes of tomato (yellow round and yellow oval) are in a 1:1 ratio.

Phenotype	Observed number	Expected number	(O – E)	(O – E)²	$\frac{(O-E)^2}{E}$
yellow round	565				
yellow oval	435				
TOTAL	1000				

Now that it's become a bit more complicated, your cousin is showing more of an interest. You suspect this is simply because he is delighted there's something you don't understand. He's always been a bit sceptical of your trust in simple genetic ratios and, in what he considers to be a moment of genius, suggests that the significant deviation from a 1:1 ratio may be the result of non-genetic factors.

Question 6: Speculate on any plausible non-genetic reasons why the data may deviate from a 1:1 ratio. (10%)

You suspect that non-genetic reasons for a deviation from 1:1 are unlikely, and try to think of a plausible genetic explanation. You remember your teacher suggesting that sometimes crosses involving multiple genes have rather strange ratios. You look at the data again and wonder whether you are actually looking at a 9:7 ratio.

Question 7: Use Table 2.9.6 to help you calculate the chi-squared value and test whether the data in Table 2.9.4 are in a 9:7 ratio. Again there is one degree of freedom and the 5% cut off is 3.84. (10%)

Table 2.9.6   Use this table to calculate the chi-squared value to test the theory that the two classes of tomato (yellow round and yellow oval) are in a 9:7 ratio.

Phenotype	Observed number	Expected number	$(O - E)$	$(O - E)^2$	$\dfrac{(O - E)^2}{E}$
yellow round	565				
yellow oval	435				
TOTAL	1000				

Question 8: Using appropriate genetic symbols, explain why the ratio of yellow round tomatoes to yellow oval tomatoes in the F2 is 9:7. (15%)

Question 9: Propose a model to explain the genetic basis of tomato shape. (15%)

Question 10: How would you test the model you proposed in your answer to question 9? (10%)

Chapter 2.10

The Curse of the WERE Rabbits

Captains of the old whaling fleets knew about the End of the World Islands and so did many of the ancient maritime explorers, but it's not a place that many of us have heard of. However, the End of the World Islands have an interesting history. They were originally discovered in the late 19th century by the largely forgotten explorer Captain Jane Doe, after a huge storm damaged her ship, *The Satin Angora*, and forced it to take refuge in the End of the World Islands' natural harbour. A few weeks later, after some hasty repairs, *The Satin Angora* set sail for home. Over the years many sailors have similar tales to tell and they all owe their lives to the natural harbour of the End of the World Islands. However, the survival of the sailors has come at some considerable cost to the End of the World Islands' endogenous and unique wildlife, as anything that is tasty has been consumed to extinction.

The Satin Angora also introduced European rabbits (*Oryctolagus cuniculus*) to the End of the World Islands, and these progenitor rabbits have evolved to form distinct populations on each of the two largest End of the World Islands, the unimaginatively named West End Island and East End Island. A wide channel of water, which has a well-deserved reputation for strong tides and currents, separates the two islands. It is generally thought that over the past 400 years the founding rabbit populations of East End Island and West End Island have evolved quite independently from one another.

The East End and West End islands have very different environments and the phenotypes of the rabbit populations on each island reflect this. East End Island is very dry and sandy (see Fig. 2.10.1) and the East End Island rabbits have golden fur and brown eyes, which presumably acts as camouflage to hide them from the endemic population of eagles. In addition, they are quite small, presumably because food on the island is relatively scarce.

Figure 2.10.1   A view of East End Island. Note the sandy environment and sparse vegetation (in contrast to the lush green vegetation of West End Island shown in Fig. 2.10.2).

West End Island has a much higher rainfall than East End Island, giving it a lush green environment (see Fig. 2.10.2). Because food is plentiful, the West End Island rabbits are much larger than their East End Island cousins. Like their European ancestors, the West End Island rabbits are grey and, but for reasons that have never been satisfactorily explained, they have evolved red eyes. As a consequence they are known as West End red-eye rabbits (or WERE rabbits for short).

Figure 2.10.2   A view of West End Island, showing the researchers campsite within the lush green landscape, with the west coast of East End Island in the distance.

Being two small distinct populations derived from the same small group of original founders at a known point in recent history, the End of the World Islands rabbits are therefore a perfect resource for an evolutionary genetic study. However, recent research has shown that the WERE rabbit population is in serious decline, although the reasons for this are unclear. As you approach the final year of your genetics degree, you are fortunate enough to get a summer placement in the laboratory of respected rabbit researcher Doctor Warren Burrows and you set about investigating the 'Curse of the WERE Rabbit'.

The decline of the WERE rabbits has been so rapid that Dr Burrows and his colleagues have had to act very quickly. They are concerned about losing the unique genetic history of the WERE rabbits, so they have set up three captive populations: one derived exclusively from WERE rabbits, one derived exclusively from East End Island rabbits, and a mixed population. Dr Burrows provides you with some data concerning the basic biology and genetics of the rabbits of the End of the World Islands. You first look at the data on body weight and these are presented in Table 2.10.1.

Table 2.10.1   Mean weights (in kilograms ± standard error) of WERE rabbits and East End Island rabbits born, captured and weighed in the wild (wild), or born and weighed in captivity (captive). The mixed rabbits born and weighed in captivity are F1 animals from pure captive WERE rabbit mothers and East End Island fathers. No F1 wild mixed rabbits exist as the West End Islands are separated from the East End Islands by a dangerous sea channel. Note that the same machine was used to weigh all rabbits. $N = 15$.

	West End	Mixed	East End
Wild	2.1 ± 0.3	NA	1.2 ± 0.2
Captive	2.4 ± 0.4	2.3 ± 0.5	2.2 ± 0.3

Question 1: Is the size difference between WERE rabbits and East End Island rabbits primarily genetic, primarily environmental or some combination of the two? How could you test your theory? (10%)

In addition the body weight data, Dr Burrows also has provisional genetic data that he hopes you will be able to interpret. When he first set up the mixed population described earlier, he crossed pure WERE rabbits (red-eyes and grey fur) to pure East End Island rabbits (brown-eyes and golden fur) to generate an F1 generation comprising rabbits that all had rich chocolate coloured eyes and grey fur. He then crossed pairs of F1 rabbits together to produce an F2 generation with the phenotypes and frequencies shown in Table 2.10.2.

Table 2.10.2   Relative ratios of F2 offspring from the crossing scheme described earlier.

Body colour	Grey	Grey	Grey	Golden	Golden	Golden
Eye colour	Chocolate	Brown	Red	Chocolate	Brown	Red
Ratio	6	3	3	2	1	1

Question 2: How are body colour and eye colour inherited in the End of the World Islands rabbits, and why do the F2 offspring occur in the ratio shown in Table 2.10.2? (15%)

Dr Burrows suggests that you start your investigations by studying the mixed population of rabbits. He is keen to establish why the WERE rabbits, a healthy specimen of which is shown in Fig. 2.10.3, are in serious decline and suggests that studies of the mixed population may reveal the answer. You are not convinced but, as it is your first week in his laboratory, you simply assume he knows what he's doing.

Figure 2.10.3   A healthy West End Red-Eye (WERE) Rabbit in the lush green pasture of West End Island.

Dr Burrows started the mixed population a year ago. He mated three true-breeding WERE rabbit does to three true-breeding East End Island rabbit bucks and then kept their offspring in a large secure outdoor arena. He asks you to capture every rabbit and record its body colour and eye colour phenotypes. After a busy day's work, you have a full set of data from all 200 rabbits. As far as you can tell, rabbit body colour is in Hardy–Weinberg equilibrium. However, the eye colour data, which Dr Burrows simply describes as 'weird', are shown in Table 2.10.3.

Table 2.10.3   The number of rabbits with red, chocolate or brown eye colour in Dr Burrows' mixed rabbit population originally derived from three WERE rabbit does and three East End Island rabbit bucks one year ago.

	Red	Chocolate	Brown
Number of rabbits	26	33	141

Question 3: What are the frequencies of the red-eye and brown-eye alleles in the mixed population of rabbits? (5%)

$$\chi^2 = \sum \frac{(O - E)^2}{E}$$

Figure 2.10.4   The chi-squared formula.

You decide to investigate whether eye-colour in the mixed population is in Hardy–Weinberg equilibrium, and remember it's got something to do with p^2, $2pq$ and q^2. The chi-squared formula is shown in Fig. 2.10.4, and for two degrees of freedom, the cut off at 0.05% is 5.99.

Question 4: Is rabbit eye colour in Hardy–Weinberg equilibrium? Speculate on why or why not. (15%)

Question 5: How would you test the hypothesis you have proposed in your answer to question 4? (10%)

Whilst you are thoroughly enjoying your time in Dr Burrow's laboratory, you are rather hoping he won't ask you to collect phenotype data again. You're not surprised to have acquired a few scratches from the nervous rabbits, but you hadn't expected so many of the evil creatures to bite you. However, it's only when you look in detail at the large spreadsheet of data that you've collected that you realise that nearly all the bites were inflicted by red-eyed and chocolate-eyed rabbits.

You spend much of the next few days observing and recording the behaviour of the three captive populations of rabbits. It soon becomes apparent that in the captive population of WERE rabbits, nearly all the animals exhibit general lethargy and fever, coupled with aggression and hydrophilia. However, these symptoms are absent in the captive population of East End Island rabbits. In the mixed population, many of the red-eyed rabbits show the symptoms of lethargy, fever, aggression and hydrophilia, whereas most of the chocolate-eyed rabbits show mild symptoms, and only a few of the brown-eyed rabbits seem to be affected.

Within days of your discovery, the laboratory is visited by Professor Anne Gorrer, the world expert on bunnypox, an infectious viral condition that is exclusively found in rabbits (see Fig. 2.10.5). Professor Gorrer has shown that the bunnypox virus infects the rabbit via a cell-surface receptor encoded by a gene called *pox-infection-entry* (*pie*). Professor Gorrer soon confirms that the sick rabbits have a bunnypox infection and suspects that the WERE rabbits have a mutation in the rabbit *pie* gene.

Figure 2.10.5   A rabbit that is sadly suffering from bunnypox, showing a fatal combination of lethargy and hydrophilia.

Question 6: How could a mutation in the rabbit *pie* cell-surface receptor gene make the host animals hypersensitive to bunnypox? Would you expect such a mutation to be dominant, recessive or somewhat intermediate in phenotype? (10%)

However, DNA sequencing of the *pie* cell-surface receptor gene in WERE rabbits and East End Island rabbits reveals that their *pie* genes have identical sequences.

You discuss your results with Dr Burrows and Professor Gorrer. You know that the WERE rabbits are highly susceptible to bunnypox, but East End Island rabbits are not. You also know that the rabbit *pie* gene, which encodes the cell-surface receptor through which the bunnypox virus enters the cell, is wild-type in both populations. In addition, in the mixed population most of the red-eyed rabbits are very susceptible to bunnypox, most chocolate-eyed rabbits less so and the brown-eyed rabbits only rarely.

Question 7: How would you use the information from your discussions with Dr Burrows and Professor Gorrer to find the gene responsible for susceptibility to bunnypox? (10%)

Question 8: Having identified a gene that, when mutant, you suspect causes susceptibility to bunnypox in the WERE rabbits, how would you prove you'd found the correct gene? (5%)

Question 9: Speculate on what type of protein this new bunnypox susceptibility gene may encode and discuss whether the mutation is dominant, recessive or intermediate. (10%)

Professor Gorrer is impressed with your work and is interested in whether you may wish to start a PhD project in her laboratory once you've finished your degree. Her laboratory is a world leader in the generation and study of transgenic rabbits in infection research. She is keen to discover whether you are the right person for her team and as part of the interview process asks you to prepare in advance an answer to the following question:

Let us suppose that we have made a transgene which comprises a wild-type copy of the new bunnypox susceptibility gene you identified earlier, and we express it under a strong control region. Let us also suppose that the transgene integrates successfully into a wild-type rabbit and expresses the wild-type copy of the new bunnypox susceptibility gene at a high level in a heterozygous rabbit. So far, so good. It later transpires that however they organise their crossing schemes, the researchers can never generate adult rabbits that are homozygous for the transgene. Suggest two reasons why this may be the case.

Question 10: How are you going to answer Professor Gorrer's question? (10%)

Advanced

The Legend of *Neptune's Cutlass*

You have just completed the final year of your Genetics degree and have been offered the opportunity to start a research PhD in the laboratory of Dr Thomas Townshend at the University of Sydney in Australia. Dr Townshend's research focuses on human colonization of the Islands of the South Pacific and for many years he has been fascinated by *The Legend of Neptune's Cutlass*, a story he now believes is true.

It was in 1780 that a fleet of 20 pirate ships, lead by the infamous Captain Handsome, set sail from Easter Island in the South Pacific planning to travel to Christmas Island in the Indian Ocean. Without warning they encountered one of the most powerful southern hemisphere storms ever recorded. Contemporary records suggest that none of the ships, nor any of the pirates, were ever seen again. However, in 1782, debris that may have been from the *Neptune's Cutlass*, Captain Handsome's ship, washed up onto the beaches of southern Fiji (see Fig. 3.1.1).

Genetics? No Problem!, First Edition. Kevin O'Dell.
© 2017 John Wiley & Sons Ltd. Published 2017 by John Wiley & Sons Ltd.

Figure 3.1.1   A contemporary model of *Neptune's Cutlass* that recently sold at auction.

Legend has it that some of the pirates survived on the mysterious August Bank Holiday Island, which allegedly lies equidistant between Easter Island and Christmas Island (see Fig. 3.1.2). The existence of August Bank Holiday Island has been disputed for many years, but Professor Townshend has recently made a startling discovery. Not only does August Bank Holiday Island exist, but it is also home to a thriving population of 250 people, all of whom seem to be of European descent!

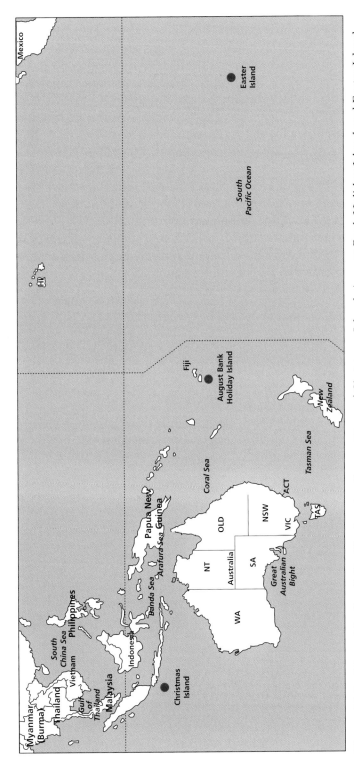

Figure 3.1.2   Map of the South Pacific showing the positions of Christmas Island, August Bank Holiday Island and Easter Island.

Dr Townshend is the first person for over 230 years to make contact with the August Bank Holiday Island population. According to the islanders, Captain Handsome managed to survive the storm by beaching the battered *Neptune's Cutlass* on August Bank Holiday Island. Indeed, it now transpires that not only did Captain Handsome and three other pirates survive, but so did their wives! All the circumstantial and historical evidence suggests that everyone on August Bank Holiday Island is descended from these eight founders.

Your PhD project is to investigate the genetic history of the population of August Bank Holiday Island. Fortunately, Dr Townshend has already acquired ethical approval for the study of the August Bank Holiday Island families and has acquired DNA samples from about 40% of the population.

Everyone on the island has one of four family names, which presumably reflects the names of the four original pirates. Apart from Captain Handsome (who was English), the survivors were Captain Lush from Wales, Captain Bellissimo from Italy and Captain Guapo from Spain.

To clarify the ethnicity of the islanders, you decide to investigate the origins of the August Bank Holiday Islanders' Y-chromosomes. Previous studies have found six short tandem repeats (STRs) on the Y-chromosome that reveal ethnic origin.

Question 1: Without resorting to sequencing the entire genome of each individual, design an experiment that would allow you to determine the lengths of the six Y-chromosome STRs. (5%)

Within a few days you determine the Y-chromosome STR haplotypes of 62 living August Bank Holiday Island males. The results of these investigations are shown in Table 3.1.1.

Table 3.1.1   Y-chromosome haplogroups and family names of 62 living August Bank Holiday Island males. The number of males with a specific family name and haplogroup are shown, along with each haplogroup's apparent ethnic origin. The six numbers within a haplogroup represent the six ordered Y-chromosome STR lengths.

Family	Males	Haplogroup	Haplotype
Handsome	10/10	**Northern Europe**	15-12-25-10-14-13
Lush	9/18	**Northern Europe**	14-12-23-11-13-13
	4/18	**Northern Europe**	14-12-23-10-13-13
	1/18	**Northern Europe**	16-12-25-13-11-13
	4/18	**Southern Europe**	18-13-24-10-15-16
Bellissimo	10/13	**Southern Europe**	18-13-24-10-15-16
	3/13	**Southern Europe**	18-15-24-12-16-16
Guapo	14/21	**Southern Europe**	18-15-24-12-16-16
	7/21	**Southern Europe**	18-13-24-10-15-16

Question 2: What does the haplotype data shown in Table 3.1.1 reveal about the paternal ancestry of the Bellissimo, Guapo, Handsome and Lush families on August Bank Holiday Island? (5%)

There is relatively little documentation regarding the original *Neptune's Cutlass* pirate families. However, in addition to being a notoriously violent man, Captain Handsome is reported to have been very arrogant and very vain. Before the fateful sailing he had commissioned a painting of his crew by the renowned 18th century portrait artist Sir Wally Bee. The painting, which is extraordinarily valuable, survives to this day and you are given special permission to study it.

The painting, which is reproduced with permission in Fig. 3.1.3, is believed to show Captain Handsome and members of his crew receiving a blessing prior to their perilous and fateful journey. Captain Handsome was well-named. He was tall, with masses of dark curly hair, an impressive beard and a patch over his left eye. The other pirates were similarly tall, dark and handsome, but all seem to have a full quota of eyes. Curiously their wives were quite unattractive, but they all appear to have a full quota of eyes.

Figure 3.1.3   Sir Wally Bee's painting of 'The Pirates and their Families'. Contemporary accounts suggest that Captain Handsome is the bearded man sitting opposite the priest.

Dr Townshend shows you several photographs taken on his most recent visit to August Bank Holiday Island. Some of the images are of the children dressed in traditional pirate costumes, celebrating 'Handsome Night' in honour of their founder. You notice that several of the children have eye patches and you presume that this is part of the ceremony celebrating the life of Captain Handsome. However, Dr Townshend reveals the true story. All the children are wearing eye-patches because they are missing an eye. Dr Townshend believes this is caused by accidents as the favourite pastime of the children is throwing spears at the August Bank Holiday Island Minipigs, which are a tasty local delicacy.

You look at several of the contemporary photographs of the young islanders, one of which is reproduced in Fig. 3.1.4, and realise that all 12 of the children wearing eye-patches have them on their left eye, just like Captain Handsome. You wonder whether there is an alternative explanation to that proposed by Dr Townshend. Surely, if the children had been involved in accidents, you would expect left and right eyes to be equally affected?

Figure 3.1.4   Lush pedigree siblings from August Bank Holiday Island showing the absent left eye (ALE) phenotype. Whilst the girl has brown-eyes, the boy is one of just three blue-eyed islanders with ALE syndrome.

Question 3: What is the probability that randomly occurring eye injuries in 12 individuals would all affect the left eye? (5%)

You wonder whether there could be an alternative, genetic explanation to the absent left eye phenotype and how a genetic condition could affect left eyes, but not right eyes, as you might expect both eyes to have an identical developmental genetic basis. Nevertheless, the more you investigate this the more you appreciate that whilst humans look symmetrical on the outside, for internal architecture that is not necessarily the case. Indeed, not only are the internal organs of the body arranged in a non-symmetrical pattern, many aspects of brain function, such as handedness, also seem to be asymmetrical.

Question 4: Briefly explain how asymmetrical patterns of gene expression may arise during development. (5%)

You explain to Professor Townshend that statistically it is extraordinarily unlikely that the eye phenotype is due to random accidents. After contacting the families, you discover that the children were in fact all born without left eyes. Further investigations reveal that about 10% of the people on the island have the 'absent left eye' phenotype.

Your time as a curious genetics student lead you to suspect that 'absent left eye' is a genetic condition. To determine whether or not this is true, you ask Professor Townshend to provide pedigree data for the largest affected family on the island. This Bellissimo pedigree is shown in Fig. 3.1.5.

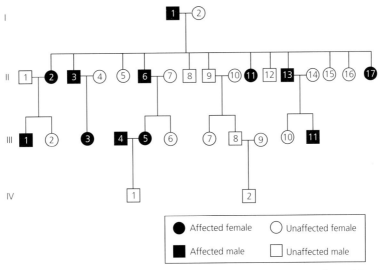

Figure 3.1.5   The Bellissimo pedigree showing individuals affected or unaffected by the absent left eye phenotype. Note that STR analyses later confirmed that this pedigree is correct.

Question 5: What is the evidence that the absent left eye phenotype is genetically inherited? What is the likeliest mode of inheritance? (5%)

Professor Townshend is very excited by your progress. In fact he is so impressed that he arranges for you to visit August Bank Holiday Island to collect more family data.

You are given a great welcome by the islanders who are extremely helpful. Very soon you notice that nearly all of the islanders affected by the condition, which you have named Absent Left Eye Syndrome (or ALE syndrome), have brown eyes. You wonder whether there could be some link between ALE syndrome and eye colour. The full set of data for all the islanders is shown in Table 3.1.2.

Table 3.1.2   Eye colour and ALE syndrome status of all 250 people living on August Bank Holiday Island.

	brown-eyes	blue-eyes
unaffected	103	122
ALE-syndrome	22	3

To determine whether there is a relationship between eye colour and ALE syndrome, you correctly decide to undertake a chi-squared test. The chi-squared formula is shown in Fig. 3.1.6. For one degree of freedom chi-squared scores above 3.84 are considered significant ($p < 0.05$).

$$\chi^2 = \Sigma \frac{(O - E)^2}{E}$$

Figure 3.1.6   The chi-squared formula.

Question 6: Use a chi-squared test to determine whether there is an association between eye colour and whether or not someone is affected by ALE syndrome. (5%)

Having collected eye colour data, you add this to the Bellissimo pedigree originally shown in Fig. 3.1.5. The modified pedigree is shown in Fig. 3.1.7.

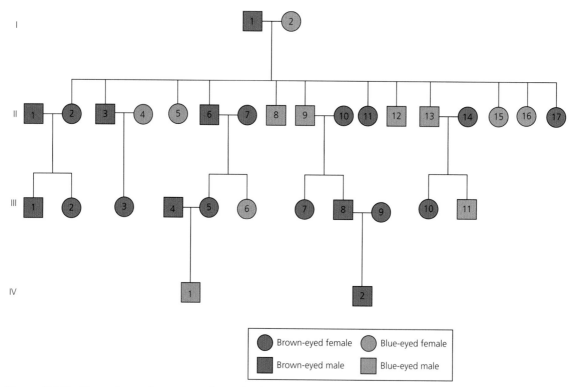

Figure 3.1.7   Eye colour phenotypes from the large Bellissimo family. This pedigree is identical to that shown in Fig. 3.1.5. Note that everyone with two eyes has eyes of the same colour: either both blue or both brown.

You look at the existing research literature and discover that approximately 70 genes determine eye colour and pattern phenotypes in human populations. However, in populations of European descent, such as those found on August Bank Holiday Island, the most important gene is *OCA2*, which is found on chromosome 15. At the *OCA2* gene, the brown allele is dominant to the blue allele.

Question 7: Redraw the Bellissimo pedigree shown in Fig. 3.1.7 and, wherever possible, annotate the pedigree with individual genotypes for the *OCA2* eye colour gene. Wherever possible, determine which individuals have inherited a parental chromosome or a recombinant chromosome. (10%)

Question 8: Calculate the approximate map distance between the *OCA2* eye colour gene and the *ALE* gene. (5%)

Dr Townshend (an unaffected blue-eyed male) is sacked for gross professional misconduct when it is discovered he is having a child with islander III:3 from the Bellissimo pedigree shown in Fig. 3.1.7.

Question 9: What is the probability that Dr Townshend and islander III:3 will have a blue-eyed daughter who is affected by ALE syndrome? (5%)

In the absence of Dr Townshend, you transfer to the laboratory of Professor Matilda Waltzing. She is fascinated by the information you have gathered. She is particularly interested that you've shown that the *ALE* gene is closely linked to the *OCA2* eye colour gene. This places the *ALE* gene quite close to *OCA2* on chromosome 15.

Professor Waltzing is an expert in state-of-the-art, high-throughput, whole-genome sequencing technologies. You acquire funding to sequence the entire genomes of two individuals and choose two sisters from the pedigree studied earlier (Figs 3.1.5 & 3.1.7). These represent one unaffected woman (II:16) and one affected woman (II:17). You focus your initial studies on the region of chromosome 15 near to *OCA2* identified in your earlier gene mapping studies. Data from this investigation is shown in Fig. 3.1.8.

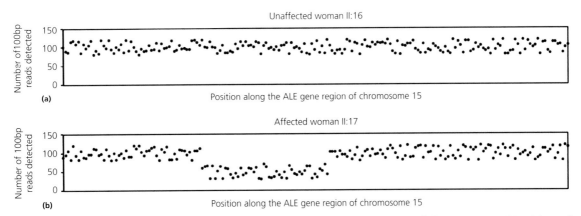

Figure 3.1.8 Frequency of genome sequence reads in the ALE gene region of chromosome 15 in: (a) unaffected sister II:16 and (b) affected sister II:17. Each dot represents a 100bp region of genomic DNA.

Question 10: Referring to the data shown in Fig. 3.1.8, what is the most plausible genetic explanation for the ALE phenotype? (5%)

Using a variety of techniques, you investigate the intron/exon structure of transcripts in the *ALE* candidate gene region of chromosome 15 described earlier. The results of your investigations are shown in Fig. 3.1.9.

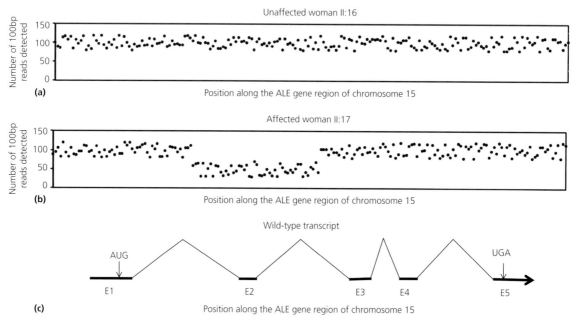

(a)

(b)

(c)

Figure 3.1.9 Frequency of genome sequence reads in the *ALE* gene region of chromosome 15 in (a) unaffected sister II:16 and (b) affected sister II:17 (as shown previously in Figure 3.1.8), with a schematic representation of the single wild-type *ALE* transcript on the same scale. The five exons, labelled E1 to E5, are shown as thick lines, whilst the four introns are shown as thinner lines. The positions of the start (AUG) and stop (UGA) codons are shown.

The wild-type mature (intron-less) *ALE* transcript shown in Fig. 3.1.9 is approximately 5.8 kb long. Unaffected individuals only produce this one transcript from the ALE gene. The sizes of the five exons (from E1 to E5) are 1591, 810, 903, 782 and 1707 bases, respectively.

Question 11: How many different *ALE* transcripts would you expect to see in affected woman II:17? What would be their approximate expected sizes and structure, and what would be their relative frequencies? (5%)

To determine ALE gene function, you enter the ALE gene sequence into BLAST. Whilst single homologues of ALE exist in all mammals, the precise function of ALE remains unresolved. Nevertheless it would appear that exon 2 has sequence similarity to a protein–protein-binding domain, and exon 3 has sequence similarity to a DNA-binding domain. This is illustrated schematically in Fig. 3.1.10.

Figure 3.1.10   Schematic representation of the single wild-type ALE transcript. The five exons, numbered E1-5, are shown as a thick line, whilst the four introns are shown as thinner lines. The positions of the start (AUG) and stop (UGA) codons are shown. Exon 2 (E2) has sequence similarity to a protein–protein-binding domain (P), whilst exon 3 (E3) has sequence similarity to a DNA-binding domain (D).

Relatively little is known about the ALE gene. However, rare deletions of the entire gene are known to exist. Individuals that are heterozygous for a wild-type copy of the gene and an ALE deletion are apparently perfectly healthy with a full quota of eyes. Therefore, you correctly conclude that deletions of the entire ALE gene are recessive.

Question 12: Propose a model that explains why the ALE mutation in the August Bank Holiday Islanders is a dominant mutation. (10%)

Question 13: How would you test the model you proposed in your answer to question 12? (5%)

You spend the next few days searching existing databases and find that a number of rare ALE mutations exist in human populations. Your career aspiration is to work as a genetic counsellor, so you are interested in predicting the consequences of being homozygous or heterozygous for each of the mutations you encounter. To help you make accurate predictions, you consider the model you suggested earlier in your answer to question 12, along with the known intron/exon structure of the ALE gene and your knowledge of different consequence of specific mutations.

Question 14: What do you think the phenotype of individuals with the following genotypes would be? (10%)
(a) Heterozygous for a wild-type copy of ALE, and a mutant ALE allele with a premature STOP codon in exon E2.

(b) Homozygous for a mutant ALE allele with the same premature STOP codon in exon E2.

(c) Homozygous for a mutant ALE allele with the same missense mutation in exon E2.

(d) Heterozygous for a wild-type copy of ALE, and a mutant ALE allele with a 2 base pair deletion in exon E3.

Your PhD is a great success. You have determined the genetic basis of the ALE mutation in the August Bank Holiday Islanders.

Following your graduation, you are keen to develop a mouse model of ALE. You contact the world authority on human eye development at the University of Paris, Professor Madame Les Yeux, and she agrees to employ you for three years.

Professor Yeux has already established that heterozygous mice carrying a wild-type ALE allele and an ALE mutation that is identical to those found in the August Bank Holiday Islanders have the Absent Left Eye mutant phenotype.

Professor Yeux also has a second mouse line with an ALE-like phenotype (see Fig. 3.1.11). However, these mice have a recessive mutation that maps to a gene on chromosome 3. She calls the gene *Oeil Gauche Absent* (OGA). The OGA gene encodes a protein with several protein–protein interaction domains.

Figure 3.1.11   Three OGA mutant mice with an ALE-like phenotype from the laboratory of Professor Yeux.

Question 15: Speculate on how the ALE and OGA genes and/or proteins may interact. (10%)

Question 16: How could you test the model you proposed in your answer to question 15? (5%)

Chapter 3.2

The Devil's Pumpkin

It is something you've never really understood. How can the world's population keep rising? Surely at some point there simply won't be enough food to feed everyone? Having just completed your genetics degree, you feel the best contribution you can make is to use your expertise in genetics to generate crop plants that can grow in harsh environmental conditions.

You are particularly interested in the potential of the crop plant the Devil's pumpkin, *Halloweenius luciferi*, which thrives in at least two distinct European populations. One such pumpkin, the Mediterranean Devil's pumpkin (*Halloweenius luciferi faliraki*) originates from the Greek island of Rhodes and is shown in Fig. 3.2.1. It is a well-characterised crop plant and is extensively used as a model genetic organism. The recently discovered Arctic Devil's pumpkin (*Halloweenius luciferi santa*) thrives in northern Finland. Its biology and genetics, especially its ability to grow in such a cold and harsh environment, are very poorly understood.

Figure 3.2.1 An orange Devil's pumpkin.

The Devil's pumpkin is of interest to researchers as the Mediterranean strain is one of the fastest growing vegetables yet discovered and can withstand the harsh environment of a typical Mediterranean summer (see Fig. 3.2.2). It is a diploid plant and its genome has been fully sequenced. It is also easy to carry out standard genetic crosses with this plant and a single mating can produce hundreds of offspring. From a plant geneticist's perspective, it has great potential.

Genetics? No Problem!, First Edition. Kevin O'Dell.
© 2017 John Wiley & Sons Ltd. Published 2017 by John Wiley & Sons Ltd.

Figure 3.2.2   A typical Mediterranean scene in summer, illustrating the harsh semi-arid environment within which Mediterranean Devil's pumpkins thrive.

The recent discovery of the Arctic Devil's pumpkin has excited researchers, as its ability to grow fast even in temperatures hovering just above freezing means that it could be developed as a major crop plant of the future in cold, harsh environments (see Fig. 3.2.3). However, relatively little is known about its biology and genetics.

Figure 3.2.3   A typical Arctic scene in summer, illustrating the harsh cold environment within which Arctic Devil's pumpkins thrive.

You have been given the opportunity to undertake a PhD with Doctor Jack O'Lantern, who is a world-renowned expert on everything related to pumpkins. You embark upon a series of experiments in his well-funded, environment-controlled, plant growth facility (see Fig. 3.2.4) with a view to understanding the genetic basis of cold tolerance in the Arctic Devil's pumpkins.

Figure 3.2.4 Dr O'Lantern's environment-controlled plant growth facility.

To acquire some basic understanding of Devil's pumpkin genetics, you begin a series of simple crosses. You start by studying the pure-breeding Arctic and Mediterranean strains as well as an F1 hybrid between the two. You grow the F1 generation at three different temperatures. The average biomass of the F1 pumpkins produced in each cross is shown in Table 3.2.1.

Table 3.2.1 Crosses were made within and between Arctic and Mediterranean strains of the Devil's pumpkin and the biomass of resulting F1 pumpkins was calculated. The average biomass score shown is the mean ± standard deviation in grammes (g) from at least 100 F1 pumpkins.

Temperature	Female gamete	Male gamete	Biomass (g)
5°C	Arctic	Arctic	10000 ± 311
5°C	Arctic	Mediterranean	9252 ± 421
5°C	Mediterranean	Arctic	9147 ± 577
5°C	Mediterranean	Mediterranean	291 ± 63
20°C	Arctic	Arctic	3011 ± 297
20°C	Arctic	Mediterranean	3225 ± 413
20°C	Mediterranean	Arctic	3464 ± 299
20°C	Mediterranean	Mediterranean	4962 ± 555
35°C	Arctic	Arctic	1111 ± 111
35°C	Arctic	Mediterranean	1561 ± 217
35°C	Mediterranean	Arctic	1743 ± 128
35°C	Mediterranean	Mediterranean	9998 ± 442

Question 1: Briefly explain what is meant by the term 'standard deviation'? (5%)

Question 2: What, if anything, do the data in Table 3.2.1 tell you about the inheritance of cold and heat tolerance in Devil's pumpkins? (5%)

It's already known that the Mediterranean Devil's pumpkin exists in two colour variants, the common orange variety and the rare white variety (see Fig. 3.2.5). This is controlled by a single gene (called *white*) that is found on the second chromosome. The orange allele (*white⁺*) is dominant to the white allele (*white⁻*).

Figure 3.2.5 A rare White Plain Mediterranean Devil's pumpkin.

In addition the Mediterranean Devil's pumpkin is usually striped, but there is a rare mutant variety called plain. The presence or absence of stripes is controlled by a single gene (called *plain*), which is also found on the second chromosome. The striped allele (*plain⁺*) is dominant to the plain allele (*plain⁻*).

The Arctic Devil's pumpkin has the same colour and pattern variants as the Mediterranean strain. You are interested in whether the genetic basis of colour and pattern in the Arctic Devil's pumpkin is the same as for its Mediterranean cousin.

Question 3: Briefly explain how you would investigate whether the genetic basis of pumpkin pattern and colour variation in the Arctic Devil's pumpkin is the same as that for the Mediterranean Devil's pumpkin. (5%)

Having established that the orange/white colour variants and striped/plain pattern variants are controlled by the same two genes in each strain, you investigate whether the genes are found in the same relative positions on the second chromosome.

According to existing data, classical genetic mapping by recombination studies show that the distance between the *white* and *plain* genes in the Mediterranean Devil's pumpkin is 28 centimorgans (map units).

You repeat the mapping study in the Arctic Devil's pumpkin.

You cross true-breeding (homozygous) plain white Arctic Devil's pumpkins to true-breeding (homozygous) striped orange Arctic Devil's pumpkins. You then perform a backcross by mating the doubly heterozygous F1 pumpkins (which have the striped orange phenotype) to the true-breeding plain white Arctic Devil's pumpkins. The numbers in each class of the four possible phenotypes are shown in Table 3.2.2.

Table 3.2.2 Frequency of offspring from a backcross between alleles at the *plain* and *white* genes of the Arctic Devil's pumpkin.

pattern	colour	F2 offspring
plain	white	347
plain	orange	146
striped	white	133
striped	orange	374

Question 4: Using appropriate genetic notation, draw the crossing scheme that enables you to map the distance between the *plain* and *white* genes in the Arctic Devil's pumpkin. (5%)

Question 5: Calculate the map distance between *plain* and *white* genes in the Arctic Devil's pumpkin. (5%)

To confirm your results, you repeat the mapping study but perform it between the two strains of Devil's pumpkin. You take a true-breeding plain white Mediterranean Devil's pumpkin and cross it to a true-breeding striped orange Arctic Devil's pumpkin. You take the F1 and backcross it to a true-breeding plain white Mediterranean Devil's pumpkin. The numbers of each class of resulting phenotype is shown in Table 3.2.3.

Table 3.2.3 Frequency of offspring from a backcross between alleles at the *plain* and *white* genes between the Mediterranean and Arctic Devil's pumpkin.

pattern	colour	F2 offspring
plain	white	475
plain	orange	21
striped	white	23
striped	orange	481

Question 6: Calculate the map distance between the *plain* and *white* genes in the cross between the Arctic and Mediterranean strains. (5%)

Using identical crossing schemes, you have now calculated the map distance between the *plain* and *white* genes from three crosses: Mediterranean × Mediterranean, Arctic × Arctic and

Mediterranean × Arctic. You had expected that in each case the map distances between the two genes would have been the approximately the same. However, this does not seem to be the case.

> **Question 7:** Speculate on why the map distances between the *plain* and *white* genes in the Arctic/Mediterranean F1 hybrid is significantly different from the map distance between the *plain* and *white* genes in the original pure-breeding Arctic and Mediterranean strains. (5%)

With a view to identifying genes that are critical for cold tolerance in Arctic Devil's pumpkins, you decide to generate mutant Devil's pumpkins using the *Agrobacterium* Ti-Plasmid. After discussions with Dr O'Lantern, you design a transgene that you hope will randomly insert into the Arctic Devil's pumpkin genome. You are fortunate that Dr O'Lantern's molecular genetics laboratory, shown in Fig. 3.2.6, is so well funded.

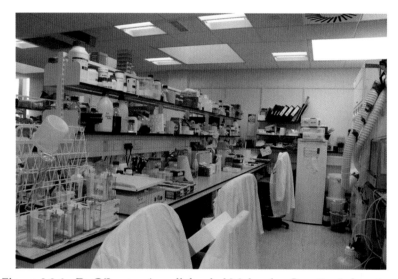

Figure 3.2.6   Dr O'Lantern's well-funded Molecular Genetics Laboratory.

You dip developing Devil's pumpkin floral tissues into a solution containing *Agrobacterium* and transfer the T-DNA into the Arctic Devil's pumpkin germ-line. The process of Devil's pumpkin transformation is shown in Fig. 3.2.7.

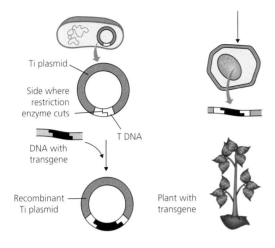

Figure 3.2.7 The process by which *Agrobacterium* T-DNA can be used to introduce foreign DNA into the germ-line of the Devil's pumpkin. According to reliable published sources, on average this process results in approximately five insertions per plant genome.

Ti plasmid

Side where restriction enzyme cuts

DNA with transgene

T DNA

Recombinant Ti plasmid

Plant with transgene

Question 8: How would you identify the DNA sequence that is mutated by insertion of the modified T-DNA? (5%)

The T-DNA you used to mutate the Arctic Devil's pumpkin contains the transgene shown in Fig. 3.2.8.

Figure 3.2.8 Schematic of the T-DNA transgene used to mutate the Arctic Devil's pumpkin. The natural T-DNA ends are shown as black bars. Between the T-DNA ends there is a *kanamycin resistance* gene (Km^R) and five directional Devil's pumpkin enhancer elements (black triangles).

Question 9: Explain the function of each of the three components of the transgene:
(a) the T-DNA ends,
(b) the *kanamycin resistance* gene, and
(c) the five directional giant pumpkin enhancer elements. (5%)

You introduce the transgene illustrated in Fig. 3.2.8 into the Arctic Devil's pumpkins and try to isolate mutants that are hypertolerant to the cold. You also introduce the same transgene into the Arctic Devil's pumpkins and try to isolate mutants that are cold-sensitive.

Molecular genetic analysis of the cold-tolerant (CT) and cold-sensitive (CS) Arctic Devil's pumpkin mutants resulting from the T-DNA transgenics, reveals that several of them seem to map to the same part of the chromosome. As this suggests that there is a critical gene for growth in cold environments, you name this gene *Frozen* (*Fro*). A cartoon of the *Frozen* gene summarizing your results and showing its structure and transgene insertion sites is shown in Fig. 3.2.9.

Further analysis of the single hyper-cold-tolerant Arctic Devil's pumpkin mutant (CT1) shows that it acts as a dominant mutation.

Further analysis of the three cold-sensitive Arctic Devil's pumpkin mutants (CS1, CS2 & CS3) shows that they are all recessive mutations.

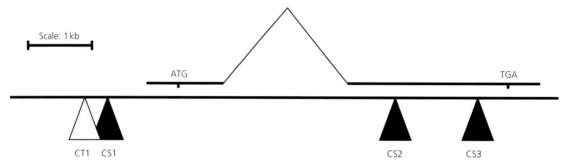

Figure 3.2.9   The molecular organisation of the *Frozen* (*Fro*) gene. The intron/exon structure is shown along with the positions of the start (ATG) and stop (TGA) codons. Molecular analysis reveals the transgene insertion site of one cold-tolerant Arctic mutant (CT1), and three cold-sensitive Arctic mutants (CS1, CS2 & CS3).

Question 10: Speculate on why the CT1 insertion in the Arctic Devil's pumpkin causes a dominant cold-tolerant phenotype. (5%)

Question 11: Speculate on why each of the three insertions (CS1, CS2 & CS3) in the Arctic Devil's pumpkin each cause recessive cold-sensitive phenotypes. (5%)

To further understand the function of the *Frozen* gene, you take three of the mutant strains described earlier (CT1, CS1 and CS2) and cross them to each other. You then study their phenotypes in the cold.

Question 12: Explain what you would expect the phenotype to be of:
(a) a CT1/CS1 heterozygous plant grown in the cold?
(b) a CT1/CS2 heterozygous plant grown in the cold? (5%)

Further analyses allow you to recover a rare recombinant where CT1 and CS1 are on the same chromosome.

Question 13: Explain what you would expect the phenotype to be of:
(a) a CT1 CS1 homozygous plant: *Fro* (CT1 CS1) / *Fro* (CT1 CS1).
(b) a CT1 CS1 heterozygote plant: *Fro* (CT1 CS1) / *Fro* (+). (5%)

After discussions with transcriptomics expert Dr Al Saints-Day, you undertake a full transcriptomic analysis of gene expression in wild-type Arctic Devil's pumpkin in the cold (5°C) and warm (35°C) to identify genes that contribute to the cold-tolerance of the Arctic Devil's pumpkin. You then calculate the relative expression of each Arctic Devil's pumpkin gene. You identify nine genes that show at least 10× higher expression at 5°C relative to 35°C. Encouragingly, the gene that is most highly upregulated at 5°C is *Frozen*. These data are shown in Table 3.2.4.

Following further discussions with Doctor Saints-Day, you repeat the experiment using the *Frozen* (CS1) mutant strain under identical environmental conditions. These data are also shown in Table 3.2.4.

Table 3.2.4   Genomic analysis of whole Devil's pumpkin plants showing relative expression levels of nine genes that are at least 10× upregulated when grown at 5°C as opposed to 35°C. The relative gene expression levels in wild-type and *Frozen* (CS1) mutants are shown ± standard deviation.

	Wild-type	Fro(CS1)
Frozen	27.2 ± 2.6	1.0 ± 0.1
Anna	16.2 ± 1.2	1.1 ± 0.2
Elsa	16.1 ± 0.6	15.7 ± 0.8
Kristoff	14.0 ± 1.1	12.0 ± 2.2
Olaf	13.7 ± 0.7	0.9 ± 0.2
Hans	12.1 ± 1.1	1.2 ± 0.4
Bulda	11.3 ± 0.9	0.8 ± 0.1
Marshmallow	11.2 ± 2.1	9.8 ± 1.6
Pabbie	10.1 ± 1.0	1.0 ± 0.2

Question 14: Using the data in Table 3.2.4, speculate on the function of the *Frozen* gene. How would you test your hypothesis? (10%)

You realise that cold is not the only stress that may affect the growth of a Devil's pumpkin, so you decide to investigate the genetic basis of drought tolerance. Using the same two strains, wild-type and *Frozen* (CS1), that were used in the previous transcriptomic analysis, you grow the Arctic Devil's pumpkins in normal and water-restricted (drought) conditions. Again, you calculate relative gene expression in normal vs drought conditions and you present these data as the relative upregulation of gene expression in the stressed (drought) environment. These data are presented in Table 3.2.5.

Table 3.2.5   Genomic analysis of whole Devil's pumpkin plants showing relative expression levels the nine genes upregulated 10x or higher in the cold described previously. The relative increase in gene expression levels in wild-type and *Frozen* (CS1) mutants in drought vs normal growth conditions are shown ± standard deviation.

	Wild-type	Fro(CS1)
Frozen	1.3 ± 0.1	1.1 ± 0.2
Anna	1.2 ± 0.3	0.9 ± 0.2
Elsa	11.1 ± 1.5	13.7 ± 1.6
Kristoff	1.2 ± 0.2	1.5 ± 0.5
Olaf	0.8 ± 0.1	1.0 ± 0.2
Hans	11.2 ± 0.8	10.9 ± 0.9
Bulda	1.0 ± 0.1	0.9 ± 0.1
Marshmallow	0.9 ± 0.2	1.1 ± 0.3
Pabbie	14.1 ± 0.9	13.0 ± 1.2

Question 15: What does the data in Table 3.2.5 tell you about the genetic basis of drought tolerance in Arctic Devil's pumpkin? (5%)

You sequence the *Frozen* gene from both the Arctic and the Mediterranean Devil's pumpkin. Analysis of the gene sequence suggests that the Frozen protein has a zinc finger domain and a protein–protein interaction domain.

  You discover eight single base pair differences between the two strains. These are listed in Table 3.2.6.

Table 3.2.6   Differences between the *Frozen* genomic sequences of the Arctic and Mediterranean Devil's pumpkin. The position of each mutant base is given as the numbers of bases upstream from the AUG start codon (-502 and -10), or as the specific codon affected (codon 9, etc.), or in the 3′ UTR. Amino acid changes that result from base pair mutations in the coding sequence are also shown.

Position	Arctic strain		Mediterranean strain	
	Base	Amino Acid	Base	Amino Acid
-502	T	not applicable	G	not applicable
-10	T	not applicable	C	not applicable
codon 9	AAA	lysine	AAG	lysine
codon 69	AGT	serine	ACT	threonine
codon 101	CTT	leucine	ATT	isoleucine
codon 175	CCT	proline	CAT	histidine
codon 201	CTG	leucine	TTG	leucine
3′UTR	G	not applicable	A	not applicable

Question 16: Speculate, giving your reasoning, on which of the eight base pair differences found between the Arctic and Mediterranean Devil's pumpkins could be the cause of the difference in cold tolerance known to be associated with the *Frozen* gene. How would you test your hypothesis? (10%)

A few weeks later two of your colleagues, Professor Candy Cane and Dr Bob Apples, are on a cycling holiday on the West Coast of Ireland when they identify what seems to be a small population of Devil's pumpkins growing in a rain-soaked field. Fortunately they have their lucky trowels with them and are able to bring living plants, which they call the Irish Devil's pumpkin, back to the laboratory.

Preliminary experiments clearly demonstrate that the Irish Devil's pumpkin belongs to the same species as the previously described Arctic and Mediterranean strains and you name it *Halloweenius luciferi leprechaun*.

All Irish Devil's pumpkins are white without stripes. They have a voracious appetite for water and demonstrate moderate cold tolerance.

Professor Cane and Dr Apples are excited about the new arrival in their laboratory because they believe this water-loving pumpkin could actually provide a route for studying drought tolerance.

Question 17: Describe in outline a series of experiments you would undertake to productively investigate drought tolerance in the Irish Devil's pumpkin. You might wish to refer to data from previous questions. (10%)

Chapter 3.3

Gravity

As a child you'd spent many a glorious summer capturing a wide variety of insects and subjecting them to robust scientific investigations. Among your more important discoveries was that insects retained in glass jars would not survive for long without food and water, especially when placed in direct sunlight. You knew this must be true as it was an experiment you'd repeated many times.

It was also clear that insects, or at least those that you'd encountered, were not particularly clever. One of your favourite experiments was to take the lid off a glass jar and turn it upside-down, then time how long it took some poor unsuspecting insect to work out that it could escape by flying down. You vividly remember one particularly angry wasp that took 27 minutes and 6 seconds to escape, but only 37 more seconds to sting your little sister (see Fig. 3.3.1). You're fairly sure this is a world record for wasp stupidity.

Figure 3.3.1   The very angry wasp before it stung your little sister.

So why do insects think escape means fly up? More to the point, how can they tell which way is up? Can they really perceive and interpret gravity?

Now, 15 years later, you're rather excited as you have the opportunity to answer this question. Having successfully completed your genetics degree, you've been given the opportunity to study the biological basis of gravitaxis in the laboratory of Dr Lightyear. He wants you to exploit the flexibility of the fruit fly *Drosophila melanogaster* (see Fig. 3.3.2) as a model genetic research organism to identify 'gravitactic genes', investigate their function and determine the molecular and genetic basis of gravitaxis.

Figure 3.3.2   Fruit flies under anaesthesia in the laboratory of Dr Lightyear.

You read the fruit fly research literature and determine a few basic facts:

- Given the opportunity, and in the absence of any other stimulus, wild-type *Drosophila* will always run or fly in an upwards direction. They are therefore said to show positive gravitaxis.
- *Drosophila* females are XX whilst males are XY.
- In gravitaxis you are measuring walking behaviour not flying ability.
- The normal life cycle of a fly is embryo, larvae, pupae, adult.

Your first task is to establish a reliable and objective system for measuring gravitactic behaviour. To achieve this, you construct the gravitaxis maze depicted in Fig. 3.3.3 from small plastic tubing.

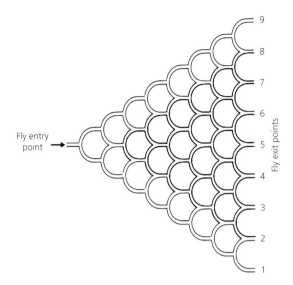

Figure 3.3.3   The gravitaxis maze. Batches of 50 flies enter the maze at the mid-point indicated by the arrow. The flies are then allowed to move through the maze until they exit at the points marked 1 to 9, where they are collected and counted.

For gravitaxis assays, the apparatus is held vertically where exit point 9 is at the top and exit point 1 is at the bottom, so flies exiting at position 9 (exit score 9) have moved up at every choice point and are therefore very positively gravitactic. Similarly, flies exiting at position 1 (exit score 1) have moved down at every choice point and are therefore very negatively gravitactic.

For control assays, the apparatus is held horizontally so that flies exiting at position 1 have turned right at each choice point, whereas flies exiting at position 9 have turned left at each choice point.

A mean exit score is simply the average exit position for a batch of 50 flies.

To establish the reliability of the gravitaxis maze, you test populations of wild-type flies and measure their response. These data are shown in Fig. 3.3.4 and Fig. 3.3.5.

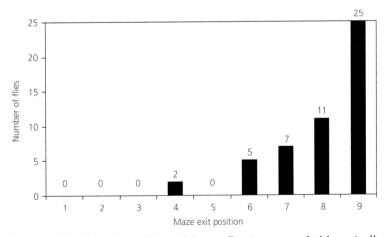

Figure 3.3.4   Behaviour of 50 wild-type flies in a maze held vertically.

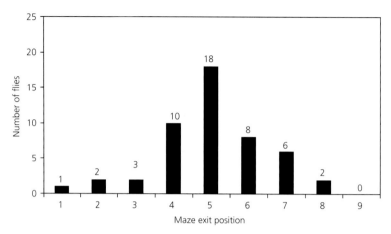

Figure 3.3.5 Behaviour of 50 wild-type flies in a maze held horizontally.

Question 1: Calculate the mean exit scores of wild-type flies when the maze is held vertically (Fig. 3.3.4) or horizontally (Fig. 3.3.5). What does this tell you about the gravitactic behaviour of wild-type flies in a maze? (5%)

As shown in Fig. 3.3.6, Dr Lightyear has a very well-funded laboratory with an extensive collection of fruit flies with which you will be able to undertake your research. Following discussions with Dr Lightyear, you decide to use the mutagen ethyl methanesulphonate (EMS) to generate gravitactic mutants. You take a wild-type population of *Drosophila melanogaster* males and give them EMS in their food. EMS will generate novel single base pair mutations in the sperm of the mutated males. You cross the mutated males to wild-type virgin females and collect their daughters. You mate the daughters to wild-type males and you collect their sons (the grandsons of the original mutated males). It is the grandsons (the F2 males) that you test for abnormal gravitactic behaviour.

Figure 3.3.6   Dr Lightyear's fruit fly laboratory.

Question 2: Using appropriate notation, draw the cross described. Show clearly how the newly mutated genes are passed through the generations and explain the rationale behind the crossing scheme, particularly why you only screen the F2 males (and not their sisters) for abnormal gravitactic behaviour. (5%)

During your screen you identify 12 gravitactic mutant males. You name these mutant males *apollo1* to *apollo12*. You cross each *apollo* mutant male to wild-type females to establish 12 stable stocks, one derived from each mutant male. It's clear that all 12 *apollo* mutations are recessive.

Question 3: How would you demonstrate that the abnormal gravitactic phenotype in each *apollo* strain is caused by an X-linked recessive mutation or a mutation on one of the autosomes? (5%)

Question 4: Having established that all 12 *apollo* mutations are on the X-chromosome, how would you determine whether they are mutations in the same or different genes? (5%)

Your results reveal that *apollo¹*, *apollo⁶* and *apollo⁹* are mutations in the same gene. More detailed analyses of these three strains suggest that *apollo¹* has a very strong phenotype, with a mean exit score of 5.0 and no evidence of gravitactic behaviour whatsoever. Both *apollo⁶* and *apollo⁹* show a weaker gravitactic phenotype, with a mean exit score of 6.5, intermediate between wild-type and the extreme non-gravitactic behaviour of *apollo¹*.

You decide to focus the rest of your research on the gene defined by the *apollo¹*, *apollo⁶* and *apollo⁹* mutations.

To determine the approximate position of the *apollo¹* mutation on the *Drosophila* X chromosome you undertake a three-point test cross.

You cross males carrying the *apollo¹* mutation, to females that are homozygous for the X-linked mutations *bald* (no bristles) and *peach* (a mutant eye colour whose wild-type allele is red).

You collect the F1 females (which must be heterozygous for all three X-linked mutations), and cross them to wild-type males. You then count the relative frequency of phenotypes in 10 000 F2 males listed in Table 3.3.1.

Table 3.3.1 Relative frequencies of phenotypes in 10000 F2 males resulting from the three-point test cross described in the text.

Bristle gene	Eye gene	Gravitaxis gene	F2
normal bristles	red eyes	gravitactic	10
normal bristles	red eyes	*apollo¹*	3 280
normal bristles	*peach* eyes	gravitactic	1675
normal bristles	*peach* eyes	*apollo¹*	25
bald	red eyes	gravitactic	30
bald	red eyes	*apollo¹*	1590
bald	*peach* eyes	gravitactic	3385
bald	*peach* eyes	*apollo¹*	5
		Total	10 000

Question 5: Draw the entire crossing scheme for the three-point test cross. Clearly show all the genotypes and phenotypes and explain the rationale behind the crossing scheme. (10%)

Question 6: Calculate the relative map distances between *apollo*, *bald* and *peach*. Draw the resulting genetic map. (10%)

Having established the approximate position of *apollo¹* on the X-chromosome, you wonder whether there are any other previously described mutations in this region that have *apollo*-like gravitactic phenotypes. You discover a recently published paper by a group of Moscow-based researchers who used a *P*-element-mutagenesis approach to identify recessive X-linked gravitactic mutants. *P*-elements are transposons that jump around the genome and can cause insertion-type mutations if they jump into and therefore disrupt a gene.

One of the mutations discovered by the Russians, *soyuz*, maps to approximately the same position on the X-chromosome as *apollo¹*. You agree to collaborate and the Russians kindly send you their *soyuz* strain. Using a complementation test you soon establish that *apollo* and *soyuz* are mutations in the same gene.

The fact that the *soyuz* mutation is associated with a transposable element has allowed the Russian group to identify the gene into which the *P*-element causing the *soyuz* mutation has inserted. They clone and sequence the *soyuz* gene and compare its sequence to that of a reference wild-type strain.

Question 7: Briefly describe how the fact that *soyuz* is associated with a *P*-element insertion makes cloning the mutant gene relatively easy. (5%)

After cloning the *soyuz* (*apollo*) gene, the Russians notice that in addition to the *P*-element insertion, the mutant *soyuz* allele has a missense mutation in its coding region that causes amino acid number 456 to change from serine to threonine (S456T).

Question 8: How would you prove that the gravitactic phenotype in *soyuz* was caused by the *P*-element insertion and not by the missense (S456T) mutation? How is it possible for a missense mutation to cause no phenotype? (5%)

You find that the *soyuz* mutant phenotype is indeed caused by the *P*-element insertion.

Now that you have established that *apollo* and *soyuz* are mutations in the same gravitactic gene, you undertake a more detailed search for fly strains carrying mutations that may disrupt the function of the *apollo* gene.

You make a detailed study of the lists of strains available from *Drosophila* stock centres. You identify five strains that contain deletions of some or all of the *apollo* gene (called deletions A–E) and a single strain that has an inversion with a breakpoint near to the 5′ end of the *apollo* gene (called inversion 1). You also identify three additional *P*-element insertions (P1–P3) that map to the *apollo* region. None of these strains have previously been tested for their gravitactic behaviour.

The molecular organisation of the *apollo* gene as well as the relative positions of the deletions, inversion and *P*-element insertions are shown in Fig. 3.3.7.

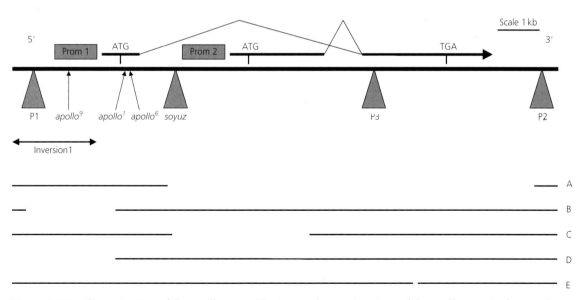

Figure 3.3.7   Organisation of the *apollo* gene. The intron/exon structure of the *apollo* gene is shown above the line. There are two promoters (Prom1 and Prom 2), two start codons (ATG) and one stop codon (TGA). The positions of the three *apollo* point mutations isolated in your screen (*apollo¹*, *apollo⁶* & *apollo⁹*) are indicated. The *soyuz P*-element insertion is shown by a triangle. The insertion sites of the three novel *P*-element insertions are shown by triangles marked P1 to P3. The limits of inversion 1 are shown by the double-headed arrow. The limits of the five deletion strains (A–E) are shown by the gaps in the line. Note that deletion D extends 15 kb to the left of the figure.

You spend the next few weeks studying the gravitactic phenotypes of each of the *apollo* mutant strains in detail. The phenotypes of each mutation in a male are listed in Table 3.3.2.

Table 3.3.2   Phenotypes of male flies carrying specific mutations in the *apollo* gene.

Mutation	Phenotype
apollo¹	no gravitaxis
apollo⁶	weakly gravitactic
apollo⁹	weakly gravitactic
P1	wild-type
P2	wild-type
P3	lethal (males die as late embryos)
soyuz (P-element)	no gravitaxis
deletion A	lethal (males die as late embryos)
deletion B	no gravitaxis
deletion C	lethal (males die as late embryos)
deletion D	lethal (males die as pupae)
deletion E	wild-type

Question 9: Referring to the information in Fig. 3.3.7 and Table 3.3.2, speculate on why males from each of the 12 strains show the indicated phenotype. (10%)

Analysis of the strain carrying inversion 1 (see Fig. 3.3.7) shows that males are slow moving. However, it's also clear that the inversion 1 males cannot perform gravitaxis, just like *apollo¹* mutant males. Curiously, heterozygous females carrying inversion 1 are also slow moving but show normal gravitactic behaviour.

Question 10: Speculate on why males carrying inversion 1 are slow moving and have no gravitactic behaviour. Similarly, speculate on why females that are heterozygous for inversion 1 are slow moving but have normal gravitactic behaviour. (5%)

To investigate the tissue-specific expression of *apollo*, you undertake northern analyses of wild-type and three mutant strains. The data depicted in Fig. 3.3.7 suggest that the gene has alternative transcriptional start sites and that you need to design a probe that will (a) identify all the transcripts indicated and (b) identify splice-specific transcripts.

Question 11: Draw a cartoon of the intron/exon structure of the *apollo* gene based on the information in Fig. 3.3.7 showing which sequence you will use as a probe for northern analysis to detect all *apollo* transcripts. In addition, indicate sequences that you could use as probes to detect each *apollo* transcript individually. (5%)

You use the probes designed in your answer to question 11 to investigate expression of the *apollo* gene in wild-type males, as well as males carrying the *apollo¹*, *apollo⁶* or *soyuz* mutations. Your results are shown in Fig. 3.3.8 (all transcripts) and Fig. 3.3.9 (transcripts from Prom 1 only).

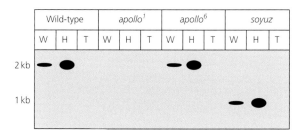

Figure 3.3.8 A schematic representation of a northern blot using a probe that should detect all *apollo* transcripts. Size markers are indicated to the left of the gel. Males from four strains were tested. Three samples were prepared for each of the four strains tested: W, whole fly; H, head; T, thorax.

Question 12: Speculate on how the EMS-induced mutations in hemizygous *apollo¹* and *apollo⁶* mutant males result in the patterns of bands shown in Fig. 3.3.8. (5%)

Figure 3.3.9 A schematic representation of a northern blot using a probe that should detect *apollo* transcripts from promoter 1 only. Size markers are indicated to the left of the gel. Hemizygous males from four strains were tested. Three samples were prepared for each of the four strains tested: W, whole fly; H, head; T, thorax.

Question 13: What is the most likely explanation for the banding pattern seen in the *soyuz* hemizygous mutant males in Fig. 3.3.8 and Fig. 3.3.9? How would you test your hypothesis? (5%)

At this point in your research you are contacted by a French group who have also undertaken an EMS screen for gravitactic mutants. One of their X-linked mutations, *ariane*, maps to the *apollo* region.

In your gravitaxis maze you find that *ariane* is weakly gravitactic, with a mean exit score of 6.5. In fact it has a very similar phenotype to *apollo⁶* and *apollo⁹*, but slightly weaker than the extreme mutant *apollo¹*.

You perform a series of crosses between the *ariane* and *apollo* mutants to generate heterozygous females. The phenotypes of the resulting females are shown in Table 3.3.3.

Table 3.3.3 Phenotypes of females that are heterozygous for two different mutations in the *apollo* gene.

Female genotype	Female phenotype
apollo¹ / apollo⁶	weakly gravitactic
apollo¹ / apollo⁹	weakly gravitactic
apollo⁶ / apollo⁹	weakly gravitactic
apollo¹ / ariane	weakly gravitactic
apollo⁶ / ariane	wild-type
apollo⁹ / ariane	weakly gravitactic

You sequence the *ariane* mutation and find that it has a C to A missense mutation in the sixth amino acid after the ATG Met codon driven by promoter 1. The *ariane* missense mutation is very close to, but not in the same codon as, the previously described mutations *apollo¹* and *apollo⁶*.

Question 14: Speculate on why *apollo⁶* / *ariane* females have a wild-type phenotype. How would you test your hypothesis? (5%)

You are approaching the end of your PhD research project, so you undertake a series of experiments to clarify the function of the *apollo* gene. The results of these experiments are listed here.

Comparative sequence analyses of the *apollo* coding regions reveals that both protein isoforms have zinc finger motifs near to their N-termini.

Transgenes containing the wild-type *apollo* transcript initiated from promotor 1 (see Fig. 3.3.7) restore gravitactic behaviour in *apollo* gravitactic mutants, but do not rescue the lethal phenotype in the *apollo* lethal mutants.

Transgenes containing the wild-type *apollo* transcript initiated from promotor 2 (see Fig. 3.3.7) rescues the lethal phenotype from *apollo* lethal mutants, but does not restore gravitactic behaviour in *apollo* gravitactic mutants.

Question 15: In light of all this information, propose a model to explain the function of the *apollo* gene. What experiments would you undertake to test your model? (15%)

Chapter 3.4

Kate and William, a Love Story

You have always been fascinated by the fictitious rodent, the Scottish haggis, *Haggis scotticus* (see Fig. 3.4.1). In the past they could be found roaming the highlands of Scotland in huge populations or clans. However, for reasons that are not entirely clear, numbers of wild haggis have fallen dramatically in recent years. The Scottish Wildlife Trust is understandably concerned that one of its great national assets is in decline and is keen to support research to understand the biological basis of that decline.

Figure 3.4.1   A Scottish haggis family basking in the autumn sun by the fence at the St Andrews Field Station. Note how their wild-type brown colouring makes them almost impossible to see.

Genetics? No Problem!, First Edition. Kevin O'Dell.
© 2017 John Wiley & Sons Ltd. Published 2017 by John Wiley & Sons Ltd.

You have been given the opportunity to undertake a PhD in the laboratory of Professor Anne Teater. She is a world authority on haggis biology. Under her supervision you plan to undertake a series of experiments to determine the cause of the decline in the haggis population and, if possible, to identify a remedy.

Prior to embarking on your PhD, you sensibly review the extensive haggis scientific literature. From your initial reading you establish that the following information is true:

- Haggis biology and genetics is essentially the same as that of mice.
- Unlike mice, haggis only give birth to one pup at a time.
- Sex determination in haggis follows the standard mammalian XX/XY system, where the Y-chromosome contains the male determining gene.

Historically, wild haggis populations in Scotland were predominantly brown, though there are fairly frequent reports in *Ye Olde Highland Herald* of rare white individuals. Estimates of haggis population size, including the numbers of brown and white haggis, are made during the St Andrews' census every June. This date is chosen as by June the population has had time to recover from the annual three-week haggis hunting season that lasts from 2nd to 23rd January (see Fig. 3.4.2). It also coincides with the haggis mating season, when courting haggis can be seen frolicking in the midsummer sunshine and are easy to count because they seem to be oblivious to passing humans.

Figure 3.4.2   Two haggis hunters in traditional dress.

Estimates of the change in the St Andrews' haggis population from 2005 to 2014 is shown in Fig. 3.4.3.

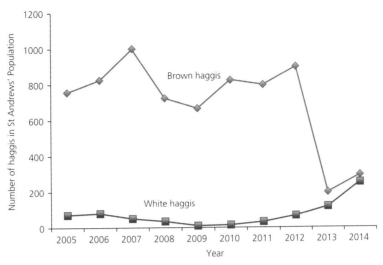

Figure 3.4.3   The number of brown and white haggis recorded in the annual (June) census over 10 consecutive years.

Question 1: January 2013 was particularly cold, with frequent deep snow recorded across Scotland, especially around St Andrews (see Fig. 3.4.4). In light of this, speculate on why there may have been such a large fall in the frequency of brown haggis between June 2012 and June 2013. How would you test your theory? (10%)

Figure 3.4.4   A white haggis family hiding in the snow next to the fence at the St Andrews' research station, after heavy snow storms in January 2013. This image is taken from the same position as that in Fig. 3.4.1.

To understand the genetic basis of body colour in haggis, you review the literature and find some data from the 1950s from Dr Jim Pansey's laboratory suggesting that rare white individuals occasionally arose due to mutations in a gene on the X chromosome.

Fortunately, Professor Teater's laboratory has a successful haggis captive breeding programme and she also has a well-established system for recording haggis family pedigrees. In fact, she has collected DNA samples from every haggis parent and pup that has passed through her laboratory.

She encourages you to investigate the inheritance of body colour in the very first haggis population established in her laboratory in the late 1990s. This first captive-bred haggis clan were derived from a pair of brown haggis caught near St Andrews. The wild-caught parents, named 'Kate' and 'William' had 12 offspring. This pedigree, including individuals from subsequent generations, is shown in Fig. 3.4.5.

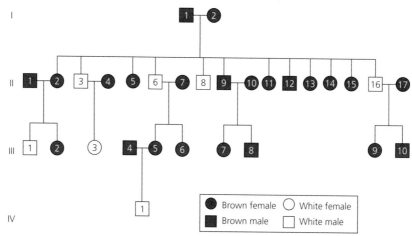

Figure 3.4.5   Pedigree analysis of the 'Kate and William' haggis family, indicating the body colour phenotype of each individual as shown.

Question 2: What is the evidence that the white body colour gene is X-linked? (5%)

To confirm definitively that the white body colour mutation is on the X chromosome, you use the stored DNA from the 'Kate and William' haggis family to determine whether the white body colour mutation is co-inherited with a well-characterised X-linked short tandem repeat (STR). The X-linked STR can vary between 6 and 15 GTA tandem trinucleotide repeats and it is not found on the Y chromosome.

You establish that the map distance between the white body colour gene and the STR is 15 map units.

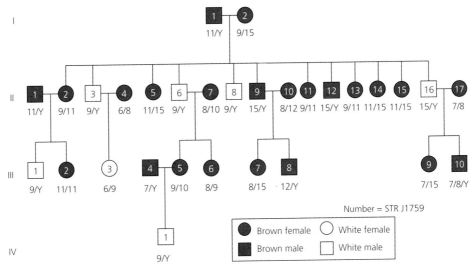

Figure 3.4.6   Pedigree analysis of the 'William and Kate' haggis family, indicating the body colour phenotype of each individual, as well as the STR genotype. Note that the pedigree in this figure is identical to that in Fig. 3.4.5 but has the STRs added.

Question 3: The evidence from Fig. 3.4.6 clearly demonstrates that the white body colour mutation and the STR are closely linked on the X chromosome. Which individual(s) show evidence of recombination? (10%)

Question 4: Speculate as to why individual III:3 is white. (5%)

Question 5: If individual III:6 is mated to an unrelated brown male with a STR genotype 14/Y, what is the probability that their firstborn pup will be a white son with STR genotype 8/Y? (5%)

The Haggis Genome Project was completed in 2008. The first haggis genome sequenced was that of 'William', the St Andrews' wild-caught haggis described earlier. Haggis have a typical mammalian genome, comprising approximately 25 000 genes.

The genome project reveals that the *white* body colour gene sits in a complex gene-dense region of the X chromosome. The organisation of the three genes is shown schematically in Fig. 3.4.7.

Figure 3.4.7   Schematic representation of the intron–exon structure of three genes, *HG43*, *white* and *wallace*, on the X chromosome of the Scottish wild haggis.

Each of the three genes in the region have been studied by various researchers. A summary of those results is given below:

- *HG43* – a gene comprising a single exon that seems to encode a protein with a DNA-binding domain. The gene is expressed early in development. The precise function of this gene – and the nature of any mutant phenotype – is currently unclear, hence its designation *HG43*, Haggis Gene 43.
- *white* – a gene comprising two exons that determines coat colour. The dominant wild-type allele encodes an enzyme that is thought to convert a white pigment to a brown pigment. This is the gene in the pedigree described earlier.
- *wallace* – a gene comprising three exons reading in the opposite direction to, and overlapping with, the *white* gene. The gene seems to encode a protein with a DNA-binding domain that is thought to play a key role in the immune response.

Most haggis populations live on mountainsides. The early highlanders noticed that haggis existed in two distinct forms: the 'lefties', in which the left legs were long and the right legs were short; and the 'righties', in which the right legs were long and the left legs short. So one form moves clockwise around the mountains and the other form moves anticlockwise around them.

Evidence from captive breeding experiments in the Haggis Genome Project supports the notion that 'lefties' and 'righties' are actually a single species.

Professor Teater gives you the opportunity to investigate the genetic basis of haggis 'footedness', which she suspects is associated with a single gene. You look at the pedigree data from all the captive-born haggis and establish several facts:

- A pair of 'righties' only ever give birth to 'righties'.
- A pair of 'lefties' always give rise to 'leftie' daughters, but some 'leftie' couples only have 'leftie' sons whereas other couples have 'leftie' sons and 'rightie' sons in equal frequencies.
- A 'rightie' mother and 'leftie' father always have 'leftie' daughters and 'rightie' sons.
- A 'leftie' mother and 'rightie' father either have all 'leftie' offspring or have 'leftie' and 'rightie' sons and daughters in equal frequencies (in other words, 25% of each).

In addition, wherever an X-linked *white* body colour mutation has spontaneously arisen in wild haggis populations, it is always co-inherited with 'footedness'.

Question 6: What is the genetic basis of 'footedness' in haggis? (5%)

DNA sequence analysis around the *white* gene reveals the molecular basis of 'footedness' in haggis. It is associated with a single base change in the *HG43* gene, from a C (leftie allele) to T (rightie allele). You rename the *HG43* gene *chirality*. This changes the sixth codon from CGA (arginine: leftie allele) to TGA (STOP: rightie allele).

Question 7: Speculate on how the C to T change in the *chirality* (*HG43*) gene affects 'footedness' in haggis? (5%)

At this point you receive some rather alarming news from the laboratory of Professor Carrie Boo, the renowned microbiologist. She has some evidence suggesting that haggis populations are actually in decline because of a virulent bacterial infection. She has identified the pathogen as *Bacillus catastrophicus*, which is shown in Fig. 3.4.8.

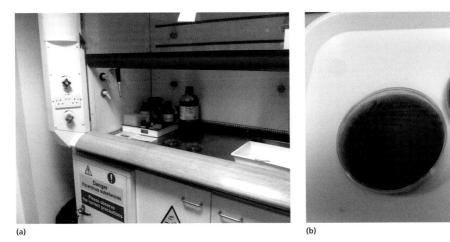

(a)   (b)

Figure 3.4.8   *Bacillus catastrophicus* growing (a) within the safety fumehood and (b) on an agar plate impregnated with haggis blood.

You are given permission by the Scottish government to use the Category Four containment facility at the University of St Andrews, shown in Fig. 3.4.9, to investigate haggis susceptibility to *Bacillus catastrophicus* infection.

Figure 3.4.9   The St Andrews' category four containment facility.

You infect some of your captive born haggis and record their survival in days. The normal life expectancy of uninfected brown or white haggis is approximately 2 years. Note that all haggis were appropriately controlled for environmental factors such as age at infection. These data are shown in Table 3.4.1. Note that no heterozygous haggis were used in these experiments.

Table 3.4.1   Survival in days of 10 homozygous brown females, 10 brown male, 10 homozygous white female, and 10 white male haggis infected with *Bacillus catastrophicus*.

Brown female	Brown male	White female	White male
3	2	49	42
2	8	36	50
6	6	43	46
4	4	53	49
9	3	45	38
5	7	47	43
7	10	42	46
7	9	45	45
3	5	47	47
5	6	44	44

Question 8: Calculate the mean (average) survival in days for each of the four classes of haggis shown in Table 3.4.1. (5%)

Question 9: What statistical test should you use to prove that white haggis survive the *Bacillus catastrophicus* infection better than brown haggis? Why is this the appropriate statistical test? (5%)

You see no logical reason, at least with regard to coat colour, why white haggis survive infection better than brown haggis. To resolve the problem, you contact Professor Pippa Strelley in the University of St Andrews' DNA Sequencing and Genomics Facility. You are given permission to capture 16 wild haggis males, eight brown and eight white, from the local St Andrews' population. You prepare a DNA sample from each of the 16 individuals and you deep-sequence their entire genomes using next generation sequencing.

You focus on the X chromosome region surrounding the *white* gene and compare the sequences of each of the 16 males to the reference sample from the first wild-caught brown male haggis 'William' described earlier. DNA sequence differences between the 16 males and 'William' are shown in Fig. 3.4.10.

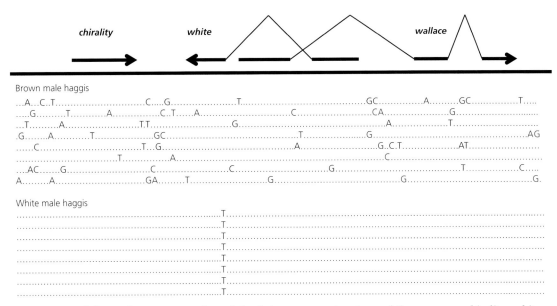

Figure 3.4.10 Schematic representation of the intron–exon structure of three genes, *chirality*, *white* and *wallace*, on the X chromosome of the Scottish wild haggis. Each horizontal line represents the DNA sequence of a specific single male. Changes in DNA sequence between each of the 16 males and the reference sample of 'William' are shown.

Question 10: Why are white haggis always relatively resistant to *Bacillus catastrophicus* infection, and why are brown haggis always relatively sensitive to *Bacillus catastrophicus* infection? (5%)

Question 11: Speculate on the evolution of *Bacillus catastrophicus* resistance in haggis. (5%)

The only gene in the white region of the haggis X chromosome that seems to function as part of the haggis immune response is *wallace*. However, according to the sequence analysis in Fig. 3.4.11, none of the resistant white haggis have a mutation within the *wallace* coding region. Therefore you decide to use qPCR to investigate *wallace* expression levels in white blood cells of white (resistant) males and brown (susceptible) males. These data are presented in Fig. 3.4.11.

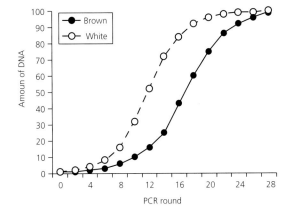

Figure 3.4.11 qPCR analysis of *wallace* gene expression in white (resistant) males and brown (susceptible) males.

Question 12: What does the qPCR analysis tell you about the genetic basis of *Bacillus catastrophicus* resistance in Scottish haggis? (5%)

So far all the experiments on *Bacillus catastrophicus* resistance have been performed on male haggis. Experiments on homozygous white females show that they are resistant to infection, whereas homozygous brown females are susceptible.

Question 13: If you exposed a heterozygous brown female to *Bacillus catastrophicus* infection, would you expect her to be resistant, susceptible or somewhat intermediate? Explain. (5%)

To identify the genes that contribute to *Bacillus catastrophicus* resistance in Scottish haggis, you undertake a full transcriptomic analysis of gene expression in white blood cells in three classes of female haggis: homozygous brown females, homozygous white females and heterozygous brown females.

You then calculate the relative expression of each of the 25 000 haggis genes. After discussions with Dr Al Packer, the local transcriptomics expert, you focus on expression of the two key genes on the X chromosome genes *white* and *wallace*. These data are shown in Table 3.4.2.

Dr Packer also suggests that you look at expression of known immune system genes. He suggests investigating four genes that he has identified as key genes in the haggis immune response against bacterial infection. These genes are called *bac1–bac4*. He also suggests a set of controls. These are four genes involved in the immune response against fungal infection called *fun1–fun4*.

Table 3.4.2   Genomic analysis of gene expression levels of 10 genes in white blood cells from three classes of haggis female, homozygous browns, homozygous whites and heterozygotes. Expression levels are given as relative levels of mRNA detected ± standard deviation. Each mean is derived from six independent samples.

		Genotypes		
		brown/brown	white/white	brown/white
Genes	*white*	40 ± 4	0	30 ± 3
	wallace	7 ± 2	43 ± 6	38 ± 8
	bac1	8 ± 4	33 ± 4	34 ± 6
	bac2	21 ± 6	83 ± 16	69 ± 14
	bac3	12 ± 5	30 ± 6	33 ± 4
	bac4	6 ± 2	9 ± 3	8 ± 1
	fun1	14 ± 6	12 ± 3	15 ± 4
	fun2	8 ± 5	43 ± 12	36 ± 7
	fun3	28 ± 7	31 ± 6	24 ± 8
	fun4	13 ± 4	10 ± 5	12 ± 4

Question 14: Taking each of the 10 genes in turn, explain why each gene has the expression levels in homozygous brown females, homozygous white females and heterozygous females shown in Table 3.4.2. (10%)

You have now completed the second year of your PhD in Professor Teater's laboratory and it is time to take stock and plan for the future.

Question 15: Briefly describe a model for how mutations in the *wallace* gene may cause resistance to *Bacillus catastrophicus* in Scottish haggis. (5%)

Question 16: You have enough funding to make genetically modified haggis in the University of St Andrews' haggis transgenic facility run by Professor Paula Bear. Design three recognisably different *wallace* transgenes that might make genetically modified haggis resistant to *Bacillus catastrophicus* and explain how each transgene works. (10%)

Chapter 3.5

The Titanians

If you hadn't studied genetics at university, you would probably have signed up for astrophysics. It is your father who strongly encourages your interest in the Solar System. He had worked as a junior researcher on the Cassini–Huygens mission and been part of the team that sent it on its amazing voyage to Saturn and its moon Titan (see Fig. 3.5.1). You remember how excited he'd been when the Huygens probe had actually landed on Titan.

Figure 3.5.1   Saturn as seen from the Cassini spacecraft.

You've always found Titan absolutely amazing. It's by far and away the largest of Saturn's moons and is actually even bigger than the planet Mercury. Titan orbits Saturn about once every 16 'earth days' and, like our own moon, its rotational period is identical to its orbital period, meaning that Titan is tidally locked in synchronous rotation with Saturn so always shows the same face to its host planet. So there is a point on Titan's surface where Saturn is always directly overhead.

You remember the date of the Titan landing, Friday 14 January 2005, very clearly. Over the next few earth weeks, earth months and earth years, your father told you about all the latest discoveries, especially progress on his main interest, finding alien life.

Images from the Cassini spacecraft and Huygens probe, revealed orange-brown sandy deserts on Titan's surface, as well as rivers, lakes and seas of liquid methane. The Huygens probe also

Genetics? No Problem!, First Edition. Kevin O'Dell.

found that Titan, like Earth, has a dense nitrogen-rich atmosphere. However, you distinctly remember your father telling you that with a surface temperature around −180°C the chances of finding some kind of complex, intelligent, methane-consuming life-form was extremely remote. However, your father was particularly excited when the Huygens probe seemed to have discovered liquid water lakes under Titan's surface, but even today astrophysicists are not entirely sure whether such lakes actually exist. However, you always felt there was something your father wasn't telling you!

Now, many years later, you've completed a very successful genetics degree and your father is wondering what your future plans are. You're fairly sure you want to pursue a career as a researcher, but you can't quite decide what aspect of genetic research excites you the most. 'Contact your mother,' he suggests. 'She works for the cheat project, and they may have a job for you.' You appreciate your father is still angry that your mother left him for a young female colleague several years ago, but are astonished that he still wants to score a few cheap points. But he soon reassures you that CHEAT is the *Cassini–Huygens Extraterrestrial Aliens from Titan* research organisation (see Fig. 3.5.2) and they have a very secret project that you, as a newly qualified geneticist, might be invited to join.

Figure 3.5.2   The impressive CHEAT Building, headquarters of the *Cassini-Huygens Extraterrestrial Aliens from Titan* research organisation.

A few days later you meet with your mother, who is absolutely delighted to see you. She explains that not only did the Huygens probe land on Titan early in 2005, but it also managed to collect significant samples from Titan's methane lakes. Initial evidence suggests that, quite incredibly, these methane samples contain living methane-consuming life! Two months ago in a remarkable covert operation designed to avoid provoking conspiracy theorists, this invaluable cargo of alien life completed its perilous journey to Earth and the aliens are now apparently alive and well in the CHEAT category six containment facility shown in Fig. 3.5.3. 'We need a geneticist to study them,' your mother tells you. 'It's your job if you want it,' she continues, 'but please be very careful, as they are the spawn of Saturn.'

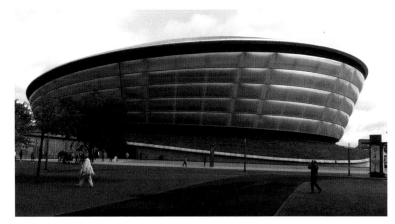

Figure 3.5.3 The CHEAT Category Six Containment Facility, originally constructed from a refurbished UFO.

On your first day at work, your mother explains in detail what the CHEAT organisation has discovered about the aliens. They have called the aliens 'Titanians' and curiously, just like life on Earth, the Titanian genetic material seems to be made of DNA. Your first task is to generate a full genome sequence of the Titanians. Although you're familiar with the new DNA sequencing technologies and suspect it will be easy to generate DNA sequence data, you're not entirely sure how you're going to build that into a full, ordered Titanian genome sequence.

Question 1: When you have acquired multiple random DNA sequences from the Titanians, explain how you could build a full, ordered genome sequence. What problems might you encounter and how might you resolve them? (5%)

Despite many problems, you eventually manage to generate an entire ordered Titanian genome sequence. Preliminary analyses reveal that Titanians have 10 independent DNA sequences, which you correctly interpret as being 10 Titanian chromosomes. The Titanians seem to be diploid with eight of these chromosomes existing in pairs, which you define as chromosomes 1 to 8. The final pair of chromosomes exists in two distinct formats, long and short, and you provisionally interpret them as sex chromosomes, though at this point you've no idea whether Titanians even have two sexes. When you compare the Titanian genome sequence with that of genome sequences from organisms on Earth, there is no obvious homology with anything.

Your mother explains why it was initially so difficult to determine that life, in the form of the Titanians, actually existed on Titan. The Titanians themselves are small and dull brown in colour, so are perfectly camouflaged against Titan's surface. In addition, it's so cold that the Titanians hardly ever move. In fact, after their initial elation at finding alien life, the researchers were a little disappointed, as not only did the aliens look bland and boring, they hardly ever did anything either.

However, in what can only be described as a moment of sheer genius, you devise a very sophisticated experiment to encourage the Titanians to do something, anything, interesting.

The experiment involved poking the Titanians with a sharp stick (see Fig. 3.5.4). At this point it is worth stressing that appropriate health and safety issues were addressed and no Titanians were seriously harmed during these experiments. The response from the Titanians was completely unexpected. The small pair of antenna positioned on the top of the head suddenly changed colour from the boring dull brown that had originally disappointed the researchers so much, to a stunning fluorescent red.

Figure 3.5.4   Sticks of the type used to arouse the Titanians were fashioned by one of the talented craftsman that work at CHEAT. You are fortunate in that CHEAT is well-funded and sticks of all shapes and sizes are available.

Over the next few weeks every available Titanian was poked with a sharp stick and every Titanian responded in exactly the same way, with fluorescing antenna. However, to the amazement of the CHEAT researchers, different Titanian populations from specific, geographically isolated underground lakes fluoresced in very distinct colours. In addition to the red-fluorescing Titanians, researchers discovered yellow-fluorescing, blue-fluorescing and green-fluorescing individuals.

As it's the only interesting feature they seem to have, you decide to determine the genetic basis, if any, of Titanian antennal fluorescence. You start by studying populations of true-breeding identically fluorescing Titanians from specific geographical regions of Titan, place them in separate supplemented cold methane environments and allow them to lay eggs. When the eggs hatch, you poke their F1 offspring with a sharp stick to reveal the colour of antennal fluorescence. The results are recorded in Table 3.5.1.

Table 3.5.1   Antennal colour of F1 sons and daughters derived from parents from four geographically isolated populations (1–4) where the populations possess fluorescing antenna of specific colours.

Population	Mother	Father	F1 daughters	F1 sons
1	blue	blue	blue	blue
2	green	green	green	green
3	red	red	red	red
4	yellow	yellow	yellow	yellow

Question 2: What does the data in Table 3.5.1 tell you about the inheritance of antennal fluorescence colour in the different Titanian populations? (5%)

You discuss your preliminary genetic data on antennal fluorescence colour with your colleague Professor Sue La Flair, who is investigating the anatomical basis of antennal fluorescence colour. She has established that the fluorescing pigment is made in a cluster of cells above the brain between the bases of the antenna and is transported up the length of the antenna where it fluoresces.

Having determined that there is a genetic basis to antennal fluorescence colour, you decide to mate different antennal-fluorescing coloured Titanians to each other. As this is a complex and time-consuming undertaking, you decide to focus your initial studies on Titanians with red or yellow fluorescing antennae. These data are shown in Table 3.5.2.

Table 3.5.2 Antennal colour of F1 offspring derived from true-breeding parents from populations 3 and 4 described in Table 3.5.1. *Note that the F1 daughters have a range of antennal colours from parental yellow through orange to parental red.

Mother	Father	F1 daughters	F1 sons
red	yellow	various*	red
yellow	red	various*	yellow

At first you are confused by the range of antennal fluorescence colours found in the F1 Titanian daughters revealed in Table 3.5.2. Each individual F1 daughter has uniform antennal colour, but different individuals may be entirely parental yellow through orange to parental red, or anything in-between. Your first impression is there are significantly more Titanian F1 daughters with orange antennae than yellow or red. To confirm this you devise a quantitative system that grades antenna colour on a 10-point scale where 0 is parental yellow, 10 is parental red and the precise intermediate orange colour scores 5. These data are shown in Fig. 3.5.5.

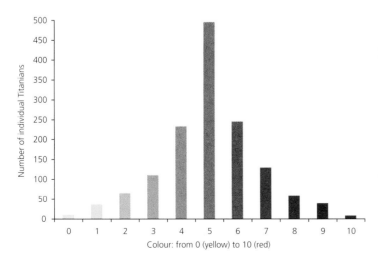

Figure 3.5.5 Number of F1 Titanian daughters with a specific colour from a cross between a red mother from population 3 and a yellow father from population 4, or *vice versa*. Colour is measured on a scale of 0 (yellow and indistinguishable from the yellow parent) to 10 (red and indistinguishable from the red parent) where 5 is orange (precisely intermediate between the parental colours).

Question 3: Propose a model that explains the genetic basis of fluorescing red and yellow antennal colour in Titanians. (10%)

To try to clarify these results, you carry out two further crosses. You take F1 females as described in Table 3.5.2 and mate them to red or yellow males from the original true-breeding strains from populations 3 (red) and 4 (yellow) described in Table 3.5.1. You repeat this several times and the results of these crosses are shown in Table 3.5.3.

Table 3.5.3   Crosses between true-breeding red Titanian males from population 3 or yellow Titanian males from population 4 described in Table 3.5.1, and the F1 'various' female Titanians described in Fig. 3.5.5 and Table 3.5.2.

Mother	Father	Daughters	Sons
F1 'various'	red	various*	50% red: 50% yellow
F1 'various'	yellow	various*	50% red: 50% yellow

You immediately observe that if the mother has an F1 'various' phenotype, the colour of the father is irrelevant to the colour of their sons. It is also clear that the actual colour of the F1 'various' females apparently has no influence on the colour of their sons or daughters. In other words, an F1 'various' mother that happens to score 8 on the colour scale (and is therefore orange-red in colour) has precisely the same spread of offspring as an F1 'various' mother that scores 1 (yellow) on the same scale. The spread of daughter colour is shown in Fig. 3.5.6a and 3.5.6b.

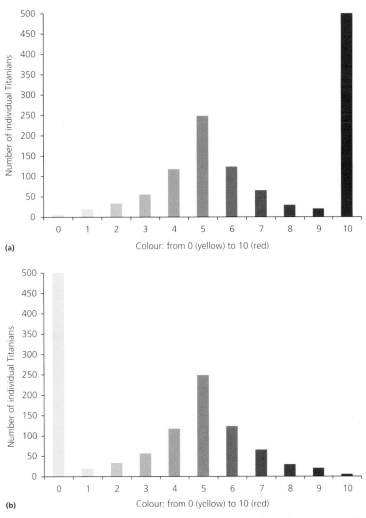

(a)

(b)

Figure 3.5.6   Number of F1 Titanian daughters with a specific colour from a cross between an F1 'vari-ous' mother (see Fig. 3.5.5 and Table 3.5.2) and a pure-breeding (a) red or (b) yellow father (see Table 3.5.1). Colour is measured on a scale of 0 (yellow and indistinguishable from the yellow parent) to 10 (red and indistinguishable from the red parent) where 5 is orange (precisely intermediate between the parental colours) as described in Fig. 3.5.5.

Question 4: Propose a model that explains the spread of colours in the daughters seen in Fig. 3.5.6. (5%)

Question 5: Speculate on what kind of protein(s) the colour gene(s) may encode and how this causes the spread of colours seen. How would you test your hypothesis? (5%)

Having resolved the genetic basis of red/yellow fluorescing antennal colour, you present your data at a top-secret meeting of all the researchers at CHEAT. It's a great opportunity to learn about other aspects of Titanian biology from your world-class colleagues. In addition to resolving the genetic basis of red/yellow antennal colour, you've also learned a lot about Titanian mating behaviour. In many respects they are the ideal model genetic organism, as they have a very short generation time. Indeed, they develop from a fertilized egg, to an egg layer in just 3 earth weeks. Titanians are just a few millimetres long and apart from their antenna, other obvious features are their feet, which comprise about half their body-weight and are essential to propel them safely through the cold and dense Titanian methane lakes. Curiously, the only way to confidently distinguish males and females is by counting their toes. Males have four toes and females just three.

Apart from the fact that male and female Titanians clearly mate, little is known about their sexual behaviour. The category six containment facility is cold and dark, just like Titan itself, and none of the researchers can actually observe quite what the male and female Titanians get up to. At the end of the day-long CHEAT researchers meeting, Professor Celeste Teal-Boddie, an expert in alien sexual behaviour, approaches you. She has been studying CCTV footage of the category six containment facility and seen some astonishing images. Whenever the facility is quiet the CCTV camera picks up pulses of fluorescing light. She has used a sophisticated algorithm to interpret the pulses. The standard pattern of light pulses from a population of red-fluorescing Titanians is presented in Fig. 3.5.7.

Time

Figure 3.5.7 Schematic representation of the pattern of light pulses exhibited by red-fluorescing Titanians. The circles represent short bursts of fluorescence, whilst the rectangles represent longer periods of fluorescence.

Question 6: If, as Professor Celeste Teal-Boddie suggests, the light pulses are an essential component of the red-fluorescing Titanians courtship ritual, what message might they be trying to get across? (5%)

To get a better understanding of Titanian courtship behaviour, Professor Teal-Boddie suggests that you look at long-term trends in mixed populations. As the genetics of red and yellow fluorescence is so complex, you decide to focus these studies on the blue and green fluorescing populations. The inheritance of the blue and green colours is associated with a single autosomal gene that shows co-dominance. Therefore homozygous Titanians can be blue or green, whilst heterozygous individuals are intermediate in colour and described rather unimaginatively as blue-green.

Professor Teal-Boddie suggests that you set up a single population initially comprising 100 blue male Titanians and 100 green female Titanians. This is defined as generation 0. You allow them to choose their own partners and collect all the eggs laid in the first four earth days. You then place the eggs in a new sealed container and allow generation 1 Titanians to hatch, breed and lay eggs, before collecting their eggs. You repeat this process every 3 earth weeks for 15 generations. The changes in frequency of the three colour types over the 15 generations are shown in Fig. 3.5.8. In generation 15, the frequency of each colour type was blue 51%, blue/green 29% and green 20%.

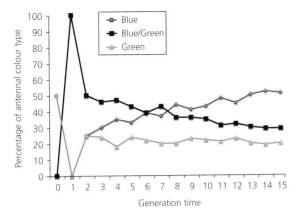

Figure 3.5.8 Frequency of the three antennal colour types in an isolated population over fifteen generations. These are representative frequencies from ten parallel experiments. Note that the percentage of blue and green individuals in generations 0, 1 and 2, are identical, being 50%, 0% and 25% respectively. In the Fig. 3.5 the green line masks the blue line.

Question 7: In generation 15, what are the frequencies of the blue and green alleles? (5%)

Question 8: In generation 15, is the Titanian population in Hardy–Weinberg equilibrium? (5%)

Question 9: What, if anything, could the frequencies of blue, green and blue/green Titanians in generation 15 tell you about the mating behaviours of blue and green Titanians? Are there any alternative explanations, and how would you test your theory? (10%)

At the next top-secret meeting of CHEAT researchers, you are approached by Professor Sharon Pluto. She is a world expert on sensory perception, specializing in mechanoreceptors. She has been investigating the function of a narrow line of bristles found on the Titanian dorsal midline. The bristles are a striking feature of the Titanians that, to the untrained eye, give the appearance of a backbone, which of course they do not possess.

At the previous meeting, Professor Pluto explained how the bristles were thought to be involved in enabling the Titanians to perceive weak gravitational forces in their cold and dark environment. She had amused the audience by presenting data on the X-linked recessive *bald* mutation. Not only do hemizygous *bald* Titanian males and homozygous *bald* Titanian females have no bristles, as a direct consequence, they have no perception of up and down. As a result even though *bald* Titanians move extraordinarily slowly, newly developed hypersensitive heat-seeking cameras were able to see *bald* Titanians endlessly crashing into each other. As a consequence, *bald* populations are very angry and hyperfluoresce.

Professor Pluto shows you some recent data on crosses that she's performed with *bald* mutants. Normally the *bald* mutation behaves as a typical X-linked recessive mutation, but in two inbred lines the data doesn't seem to follow this mode of inheritance. These data are reproducible and thus not an error in the crossing scheme. She hopes that your expertise as a geneticist will help resolve the issue. Appropriately, Professor Pluto calls the inbred strains Confused-1 and Confused-2. Her original data showing a typical X-linked inheritance pattern of the *bald* mutation are shown in Table 3.5.4.

Table 3.5.4 Crosses illustrating X-linked recessive inheritance of the *bald* mutation.

Mother	Father	Daughters	Sons
wild-type (+/+)	*bald* (−/Y)	wild-type (+/−)	wild-type (+/Y)
bald (−/−)	wild-type (+/Y)	wild−type (+/−)	*bald* (−Y)
wild-type (+/−)	wild-type (+/Y)	50% wild-type (+/+)	50% wild-type (+/Y)
		50% wild-type (+/−)	50 *bald* (−/Y)

In the two Confused strains, crosses between specific pairs of *bald* females (−/−) and wild-type males (+/Y) sometimes generate an F1 comprising a phenotypic ratio of 3 wild-type females, 1 wild-type male and 4 *bald* males. Further tests reveal the phenotypically wild-type males from this cross are sterile. Your first thought is that Professor Pluto has stumbled across an autosomal recessive sex-determining mutation, where the class of phenotypically wild-type males are chromosomally XX.

Question 10: Draw an annotated crossing scheme for the mating between a *bald* female and wild-type male that supports Professor Pluto's hypothesis that the Confused-1 and Confused-2 strains have a mutation in an autosomal recessive sex-determining gene. (10%)

On hearing about your progress, Professor Teal-Boddie is desperate to talk to you. She has been looking at the genetic basis of Titanian sex-determination using a Next Gen RNA-seq approach. She has identified over 100 Titanian genes that show sex-specific differences in expression levels (mRNA quantity) or are sex-specifically spliced (mRNA quality). You undertake Next Gen RNA-seq of the Confused-1 and Confused-2 mutant strains and search for any abnormalities in expression levels or splicing patterns in genes previously identified by Professor Teal-Boddie as potential sex-determining genes. Next Gen RNA-seq analysis of one particularly interesting autosomal gene is shown in Fig. 3.5.9.

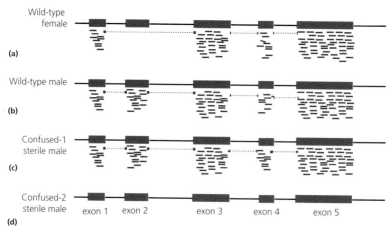

Figure 3.5.9 Next Gen RNA-seq data from a gene identified by Professor Teal-Boddie as being alternatively spliced in (a) wild-type female and (b) wild-type male Titanians. Next Gen RNA-seq data from the same gene in sterile male Titanians (which are homozygous for an autosomal sex-determining gene mutation) in the (c) Confused-1 and (d) Confused-2 strains. The black line represents the DNA sequence of the region as determined by the Titanian genome project. The predicted exon structure of the gene is shown as orange boxes within the genomic DNA sequence. Representative 100nt mRNA reads are shown in blue.

Your first thought is that this cannot possibly be the sex-determining gene as the RNA-sequencing data suggests the Confused-1 sterile mutant males produce a perfectly normal transcript. But then you remember that no one has actually proved that Confused-1 and Confused-2 sterile mutant male have mutations in the same gene.

Question 11: Without resorting to any molecular genetic analyses, how could you prove that Confused-1 and Confused-2 sterile mutant males have mutations in the same gene? (5%)

Your results confirm that Confused-1 and Confused-2 have mutations in the same gene, so you look again the data in Fig. 3.5.9. You are convinced that the *confused* gene plays a key role in sex determination and, as the CHEAT genetics expert, you know that sooner or later you are going to have to explain the role of this gene to your mother.

Question 12: Speculate on how the gene is alternatively spliced in males and females. (5%)

To investigate this further, you contact CHEATS expert biochemist, Professor Stella Rotation. She helps you investigate the sex-specific nature of the Confused protein using a Western Blot. This data is shown in Fig. 3.5.10.

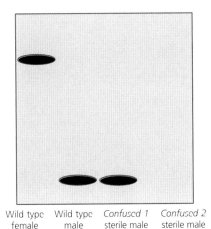

Wild type female   Wild type male   *Confused 1* sterile male   *Confused 2* sterile male

Figure 3.5.10   Western Blot of the Confused protein in wild-type female, wild-type male, and sterile males that are homozygous for either the *confused-1* or *confused-2* mutation. Higher molecular weight proteins are at the top of the image.

Question 13: According to the data presented in the Western Blot in Fig. 3.5.10, what part of the Confused protein might the antibody targeted at? (5%)

Question 14: Speculate on how the *confused* gene might determine sex in Titanians? (10%)

Question 15: If the *confused* gene encoded a transcription factor, how might it determine sex? How would you test this theory experimentally? (10%)

Chapter 3.6

Once Bitten, Twice Shy

Not only is the Zombie Institute for Theoretical Studies (ZITS) the world's only research facility dedicated exclusively to the scientific study of zombies, it is also the world's leading research facility dedicated to the scientific study of zombies. The institute is housed at the University of Glasgow, within the classical style building shown in Fig. 3.6.1.

Figure 3.6.1 The Zombie Institute of Theoretical Studies building at the University of Glasgow.

You have dreamed of being a member of the ZITS research team ever since one of their most famous researchers, the Theoretical Zombiologist Professor Leesa Lane, visited your school as part of the ZITS Public Engagement programme. Indeed, this was such an important part of your childhood that a poster of the ZITS logo is still fastened to your bedroom wall (see Fig. 3.6.2). For as long as you can remember, you've been a regular visitor to the

Genetics? No Problem!, First Edition. Kevin O'Dell.
© 2017 John Wiley & Sons Ltd. Published 2017 by John Wiley & Sons Ltd.

ZITS website (http://www.zonbiescience.org.uk), which has regular updates on their research and public education activities, as well as the latest news on alleged sightings of zombies.

By a remarkable stroke of good fortune you recently discovered that Professor Lane and her renowned team of zombie experts were looking for a PhD student to strengthen their zombie genetics research team and you were delighted when they gave you the opportunity to work with them.

Figure 3.6.2   The Zombie Institute for Theoretical Studies logo.

You've always been interested in zombies. Indeed, you think it's something that must be in your blood. In the 1940s your great uncle was a famous zombie-hunter and many of your favourite childhood memories were listening to stories of his exploits. He was clearly a very brave man and a bit of a hero of yours, even though you suspect that most of the stories of his adventures were considerably embellished (see Fig. 3.6.3). Nevertheless, it still amazes you that your great uncle survived all the dangerous encounters.

Figure 3.6.3   An image from the collection of your great uncle, the famous zombie-hunter, showing his colleague inspecting the carcass of a moose that he suspected may have been killed in a zombie attack.

On your first day at ZITS, one of Professor Lane's most experienced colleagues, Dr Smith (see Fig. 3.6.4), explains that he has identified a family exhibiting what he believes is a form of mild, inherited zombieism.

Figure 3.6.4   Dr Smith, one of the most experienced ZITS researchers.

Dr Smith explains that several generations of the family apparently suffering from the mild, inherited form of zombieism, are affected by a late-onset, progressive neurodegenerative-like condition that is characterised by an unsteady shuffling walk, poor speech that progresses to an incessant moan, and an insatiable desire for food. Dr Smith speculates that the family suffers from an inherited autosomal dominant single-gene disorder that he describes as 'mild late-onset zombieism'. An image of the pedigree of the affected family is reproduced here in Fig. 3.6.5.

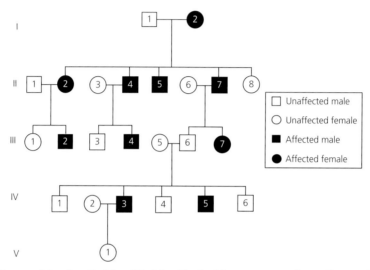

Figure 3.6.5   Pedigree of the family identified by Dr Smith as apparently suffering from an autosomal dominant, mild, late-onset zombie-like condition.

Question 1: What is the evidence supporting Dr Smith's hypothesis that 'mild late-onset zombieism' within the pedigree shown in Fig. 3.6.5 is an autosomal dominant inherited condition? Are there any problems with this hypothesis? (5%)

Question 2: What, if any, is the significance that individual III:6 died at 29 years of age in an incident that involved three as yet unidentified individuals, a chain saw and an oven set at 210°C? (5%)

You are very excited about Dr Smith's findings and ask whether it has been possible to collect DNA samples from each of the individuals in the affected pedigree. Dr Smith points out that, not surprisingly, many of the people in the pedigree are actually dead (and very possibly undead), but that he has acquired permission for tissue samples to be stored under strict conditions in a −80°C freezer within the containment facility at ZITS.

Using the state-of-the-art facilities at ZITS, you generate whole genome DNA sequences from each family member. Note that although there are 24 people in the pedigree, the DNA sample of one individual, III:6, is unavailable as DNA will not survive a temperature of 210°C. Two weeks later you have the whole genome DNA sequence of 23 members of the family identified by Dr Smith.

Question 3: How would you use the DNA sequence data from the 23 completely and accurately sequenced genomes to identify the mutation that causes the mild-late-onset, zombie-like phenotype identified by Dr Smith? What problems might you encounter? (10%)

Question 4: If you could check one base pair of DNA sequence per second, how long in years would it take you to check the entire six billion (6 000 000 000) base pairs of a diploid human genome? (5%)

Preliminary investigations suggest that the mild-late-onset, zombie-like condition is associated with a D178N (aspartic acid to asparagine at amino acid 178) missense mutation in a novel gene that shows sequence similarity to a known prion protein gene. You call it the *Zombie-Associated Prion Protein Early Death* gene or *Zapped*.

A few weeks later another of the ZITS researchers, Dr Ken Howe (see Fig. 3.6.6), apparently discovers further intriguing evidence of zombie-like symptoms existing within a family.

Figure 3.6.6   Dr Ken Howe, senior ZITS researcher.

The family identified by Dr Howe is part of a complex religious cult living in almost complete isolation. Their pedigree is shown in Fig. 3.6.7. Dr Howe believes this family exhibits a rare, severe, autosomal recessive, early-onset inherited form of zombieism.

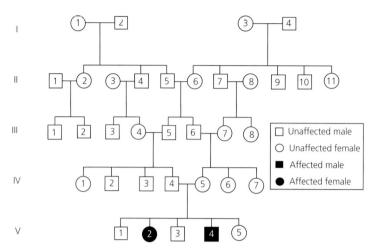

Figure 3.6.7   The pedigree identified by Dr Howe showing the individuals apparently suffering from a rare, severe, autosomal recessive, early-onset inherited form of zombieism.

> **Question 5:** What is the evidence supporting Dr Howe's assertion that the family shown in Fig. 3.6.7 has a rare, severe, autosomal recessive, early-onset inherited form of zombieism? Why might this particular pedigree reveal an autosomal recessive genetic condition? (5%)

Dr Howe encourages you to sequence the genomes of all members of the third, fourth and fifth generation of the affected family (III, IV and V in the pedigree in Fig. 3.6.7). As you hoped and perhaps expected, affected individuals V:2 and V:4 both have a missense mutation in the *Zapped* gene identified earlier. However, their shared mutation is T193I (threonine to isoleucine at amino acid 193), different from the D178M mutation segregating in the pedigree identified by Dr Smith. In addition, both V:2 and V:4 are heterozygous for the T193I mutation and no other members of the pedigree identified by Dr Howe have the T193I mutation.

> **Question 6:** What is the most plausible explanation for the inheritance pattern of the T193I mutation in the pedigree identified by Dr Howe? (5%)

A few months later you arrive at work at the ZITS building as usual and notice a black van parked outside with the unmistakable logo of the Scottish Anti-Terrorism Activity Network (SATAN). You wonder whether one of the institute's 'special' experiments has gone horribly wrong, but are at least partially reassured when you see all your colleagues in deep discussion with the men in black. It seems that there has been an outbreak of zombieism on the university campus and SATAN is keen for the ZITS team to investigate.

You realise this is a chance to shine and start by collecting as much information about the affected individuals and their close unaffected friends as possible. It soon becomes apparent that none of the affected individuals are closely related, so you speculate that there is an environmental cause for the outbreak.

Table 3.6.1   Recent experiences of students affected by zombie-like symptoms (final column), and the equivalent experiences of their close friends (other columns). +, student experienced the named behaviour; –, student did not experience the named behaviour. Note that this is only a very small subsection of the original data, which included all 53 affected individuals and hundreds of their unaffected friends, and included 43 questions.

Student	Recent activity or behaviour					
	Lots of alcohol	Refectory beef burger	Sexy business	Attended lecture	Busking on a train	Zombie-like symptoms
1	+	+	+	–	+	+
2	+	–	+	–	–	–
3	+	+	–	+	+	+
4	+	–	–	–	–	–
5	+	–	–	–	+	–
6	+	–	+	–	–	–
7	–	+	–	–	–	+
8	+	+	–	–	–	+
9	+	–	–	–	–	–
10	+	+	+	+	–	+

Question 7: According to the data in Table 3.6.1, what is the most likely source of the apparent zombieism outbreak? (5%)

You sequence the *Zapped* gene in all 23 affected individuals and find that none of them have mutations in the *Zapped* gene.

Question 8: If none of the affected students have mutations in their *Zapped* gene, explain why they are apparently showing precisely the same symptoms as those individuals that do have mutations in their *Zapped* gene. (5%)

One afternoon Dr Smith calls you into his office to tell you some rather exciting news. He has received permission to send one member of the ZITS staff to the Caribbean Island of Herspaniola. You remember hearing about Herspaniola and how zombieism was once thought to be endemic on the island. Previous research by one of the institute's former students, Cat Atonia, seems to suggest that Herspaniola zombieism is associated with cannibalism. However, that research was never completed as Cat suddenly disappeared during a lunch engagement on the island. Before she disappeared, Cat made a startling discovery. Apparently, despite continuing cannibalism, fewer Herspaniolans are showing signs of zombieism. Dr Smith is keen for you to establish why this is the case and you are keen to visit Herspaniola for as short a period as possible.

You work tirelessly during your week in Herspaniola. You manage to acquire tissue samples from 2500 dead Herspaniolans and are able to bring them back to the laboratory for DNA sequencing. Interestingly, within the Herspaniolan population you find two novel polymorphisms in the *Zapped* gene described earlier. Intriguingly these two polymorphisms haven't been found in any other populations. The polymorphisms are G127V (glycine to valine at amino acid 127) and M129V (methionine to valine at amino acid 129).

You focus your initial studies on the more common G127V polymorphism. Pulling all the data together you establish that:

- of 378 heterozygous V127/G127 individuals, 13 died of zombieism,
- of 2122 homozygous G127/G127 individuals, 401 died of zombieism.

You wonder whether the *Zapped* V127 variant confers partial resistance to zombieism.

You correctly decide to test your theory statistically with a chi-squared test. The chi-squared formula is shown in Fig. 3.6.8. For one degree of freedom, the 5% significance limit is 3.84.

$$\chi^2 = \sum \frac{(O-E)^2}{E}$$

Figure 3.6.8   The chi-squared formula.

Question 9: Use a chi-squared test to determine whether V127/G127 heterozygotes are more likely to survive zombieism than G127/G127 homozygotes. What, if anything, does that tell you? (10%)

You are interested in discovering when and where the two mutant alleles (V127 and V129) arose on Herspaniola. A map of the island of Herspaniola is shown in Fig. 3.6.9. The relative percentages of the *Zapped* polymorphic alleles along the road from Princess Port to Saint Dominica are shown in Table 3.6.2.

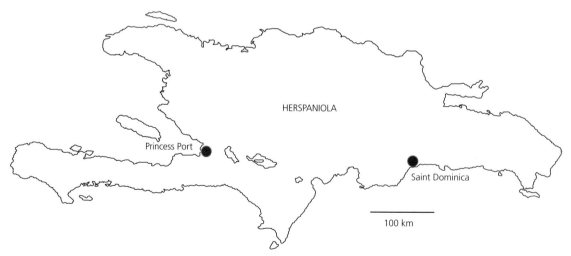

Figure 3.6.9   Herspaniola Island showing the position of the two biggest cities, Princess Port and Saint Dominica, which lie precisely east/west of each other and are approximately 320 km apart.

Table 3.6.2   Relative percentages of the *Zapped* polymorphic alleles at amino acid 127 (G127 and V127) and at amino acid 129 (M129 and V129) on the road from Princess Port (PP) to Saint Dominica (SD).

Allele	PP 0 km	40 km	80 km	120 km	160 km	200 km	240 km	280 km	SD 320 km
G127	85	86	90	92	95	98	99	100	100
V127	15	14	10	8	5	2	1	0	0
M129	99	97	92	89	84	78	75	73	77
V129	1	3	8	11	16	22	25	27	23

Question 10: Where on the road from Princess Port to Saint Dominica is each of the derived alleles (V127 and V129) likely to have arisen? (5%)

Question 11: How could you determine approximately when in history each of the derived alleles (V127 and V129) first appeared in the population? (5%)

Both the V127 and V129 variants are at small but significant frequencies in the Herspaniola population, and individuals that are transheterozygous for both alleles (genotype: G127 V129 / V127 M129), whilst not common, have been found. However, you have yet to find anyone that is homozygous for both derived alleles (V127 V129 / V127 V129). Dr Howe tells you that he's not at all surprised at that.

Question 12: Give two plausible explanations why there do not seem to be any individuals that are homozygous for both derived alleles (V127 V129 / V127 V129). (5%)

Your work has been a great success and the ZITS staff are so impressed they organise a party in recognition of the progress you've made in helping them understand the genetic basis of zombieism. It is one of the best parties you've ever attended, with industrial quantities of high-quality cake. You are just beginning to believe it really can't get any better than this when you hear the first screams. Someone, and it was never quite established who, had decided to use the containment facility to help them win a game of hide and seek and in the excitement, they'd forgotten to close the door! The screams became louder and louder and the last thing you remember is a hideous 'creature' biting your neck.

Three weeks later you're surprised to find yourself waking up in a hospital bed with a concerned looking Dr Howe sitting staring at you. As soon as he realises you're awake, a broad smile crosses his face. He explains that 16 members of the ZITS team were bitten that fateful night and you are the only victim who seems to be unaffected by zombieism. You are horrified by the CCTV footage and other images of that dreadful night (see Fig. 3.6.10).

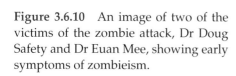
Figure 3.6.10   An image of two of the victims of the zombie attack, Dr Doug Safety and Dr Euan Mee, showing early symptoms of zombieism.

In your rather dazed state you're not at all surprised when Dr Howe asks to take a blood sample of yours back to the laboratory.

Question 13: Why might you be the only unaffected victim and why is Dr Howe so keen to take your blood sample back to the laboratory? (5%)

Even though your PhD research has been a great success and Dr Howe is delighted with your progress, you can't help feeling a little disappointed, and not just because you've lost so many of your colleagues in the 'incident'. Clearly finding the *Zapped* gene has been a key success, but your goal would be to develop treatments or cures for those unfortunate individuals that are affected by zombieism.

Question 14: What is the difference between a treatment and a cure? (5%)

Knowing that you're keen to continue your research after your PhD, Dr Howe suggests you talk again to Dr Smith, as he is an expert in genetic modification and gene therapy. Dr Smith agrees to employ you to help design and create a transgenic (genetically modified) zombie mouse model to enable you to pursue your research goals. You know the basics of how to create genetically modified mice from your undergraduate days as a genetics student, and now you have the opportunity to put your knowledge into practice. You are pleased to discover that mice have a homologue of the *Zapped* gene.

Dr Smith suggests that you first generate a strain of genetically modified mice. This transgenic mouse should replicate the original human zombie-like phenotype as closely as possible. Once that has been successfully achieved, you should then try to treat and/or cure those mice affected by zombieism using a gene therapy approach.

Question 15: Draw an annotated image of the construct you would use to allow you to make a mouse model of zombieism. Explain the function of each component of the construct, and discuss what problems you might you encounter and how might you resolve them. (10%)

Question 16: Having created a mouse model of zombieism, explain how a gene therapy approach might enable you to develop a treatment or cure for zombieism. What problems might you encounter and how might you resolve them? (10%)

Chapter 3.7

Red-Crested Dragons of Mythological Island

It is the summer after the successful completion of your genetics degree, and you're on a flight to Mythological Island (see Fig. 3.7.1) to start your PhD with Professor St George. Dragons are your passion, which is why you applied for the position in his research institute in the first place. You were pleased to be interviewed and later short-listed as 'first reserve' behind prize-winning student Sabra de Silene. You were surprised yet delighted when Professor St George phoned to tell you the position had suddenly become available again. You're excitement was only slightly tempered by the news of Sabra's serious injury, but at least Professor St George expects her to make a full recovery.

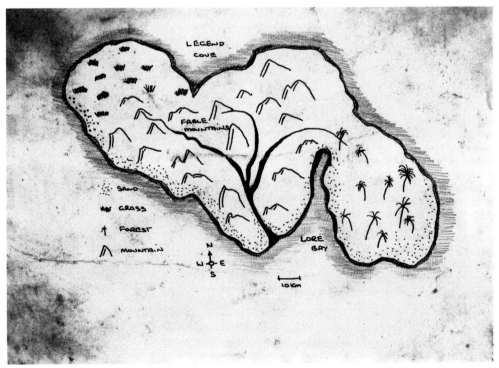

Figure 3.7.1   The earliest known map of Mythological Island dating from 1777.

Genetics? No Problem!, First Edition. Kevin O'Dell.
© 2017 John Wiley & Sons Ltd. Published 2017 by John Wiley & Sons Ltd.

As your flight approaches Deeply Shrouded Airport, you are rather nervous. The airport, which is situated in the north-east of the island in the shadow of Fable Mountain, has a rather mixed safety record. Late summer on Mythological Island marks the middle of the short but intense red-crested dragon breeding season, when males form leks and try to impress the females by competing to see which male can breathe fire the furthest. On a bad day the resulting smoke can reduce visibility at the airport to just 50 metres. Fortunately today is one of the wettest summer days in recent history and the red-crested dragons are hiding from the rain. After a few nervous moments, you are relieved when your plane lands safely.

Professor St George is at the airport to greet you. As a welcoming gift she gives you a framed copy of the oldest known image of a red-crested dragon, a copy of a cave painting of a flying female drawn by one of the earliest human settlers on Mythological Island (see Fig. 3.7.2). She explains that your research project is funded by the International Union for the Conservation of Nature (IUCN). They are very concerned about the recent, very sudden and as yet unexplained decline of the red-crested dragons (*Draconis ruficrista*) that are endemic to Mythological Island. Professor St George's research team is trying to identify the cause of the population decline and hoping to develop a plan to prevent red-crested dragons from becoming extinct.

Figure 3.7.2   A copy of a cave painting of a female red-crested dragon reproduced with permission.

You spend much of your first week studying wild populations of red-crested dragons from a very safe distance. You soon realise that the name 'red-crested dragon' isn't really appropriate as all females have small green crests and only the males have coloured crests, which may be red, brown or green (see Fig. 3.7.3). Apart from the red or brown crest colour of some males, the male and female dragons are entirely dull green. The dull green colour is thought to act as camouflage within the forests of Mythological Island, where it presumably protects them against predation by the local sub-species of the great bustard (*Otis tarda mythologica*), which, at 20 kg, is the world's largest flying bird.

Figure 3.7.3 Variation in the crest colour of male and female red-crested dragons as drawn by prize-winning student Sabra de Silene.

Professor St George explains more precisely how she wants you to approach your research project. Whilst your main focus is investigating why the population of red-crested dragons is declining in number, she's keen to get a broader understanding of the genetics and biology of the species as relatively little is known about them. It is fortunate that the Red-Crested Dragon Genome Project has just been completed.

According to the most authoritative dragon website, wikimania, in addition to the red-crested dragons of Mythological Island, there are eight other known and well-documented species of dragon. Recent comparative DNA sequencing analyses have revealed the pattern of evolutionary relationships between the eight species, and this is shown in Fig. 3.7.4. The DNA percentage similarity scores are derived from comparative analyses of the DNA sequences from 1000 genes from each of the eight extant species against homologous DNA sequences from the red-crested dragons of Mythological Island. Quite where the red-crested dragon fits into this pattern has previously been unclear, but now that the Red-Crested Dragon Genome Project has been completed, Professor St George is hoping you can resolve this question.

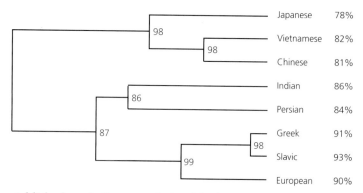

Figure 3.7.4 The established evolutionary relationship between the eight known species of extant dragon. The position of the red-crested dragon of Mythological Island is currently unknown. The percentage DNA similarity between each of the eight known dragon species and the red-crested dragon of Mythological Island is shown in black to the right of the figure. The red figures at the nodes are the result of comparative statistical analyses and represent the probability that a node is correctly positioned.

Question 1: Briefly describe the evolutionary relationship between the eight previously described dragon species shown in Fig. 3.7.4? (5%)

Question 2: Redraw Fig. 3.7.4 to show the evolutionary relationship between the red-crested dragons of Mythological Island and the other eight dragon species. (5%)

You are curious why male red-crested dragons have variable crest colours and decide to investigate this further. Professor St George suggests you study dragon mating behaviour. She has spent many hours watching fire-breathing males (see Fig. 3.7.5), and whilst females certainly seem to prefer males that breathe fire the furthest, she's sure it's a lot more complicated than that, especially as competing males are frequently seen wiggling and displaying their crests. Professor St George suggests you start by studying mating behaviour in captive-bred red-crested dragons.

Figure 3.7.5 A red-crested dragon breathing fire. Again this is a copy of an original cave painting and is reproduced with permission here.

For safety reasons, captive red-crested dragons are fed a special diet called 'Quench', which is rich in aloe vera and significantly reduces the quantity and intensity of the fire they produce. However, as captive males cannot form competitive fire-breathing leks, captive males and females are unlikely to mate. Therefore, to encourage mating in captive red-crested dragons, it is essential to use smoke machines, play the recorded deafening roars of fire-breathing lekking males through a sophisticated amplification system, and introduce the smell of rotting cooking flesh. Fortunately, Professor St George had been the drummer in a student heavy metal band in the 1970s and she knew exactly what to do.

You set up a series of female choice experiments. In each case you take a single receptive female and you simultaneously introduce her to three males, one red-crested, one brown-crested and one green-crested. You watch their behaviour and identify which of the three males she mates with. The results are shown in Table 3.7.1.

Table 3.7.1   Number of successful competitive mating's between female dragons and male dragons with red, brown or green crests.

		Female
Male	red-crested	69
	brown-crested	22
	green-crested	9

Question 3: If, as the evidence from Table 3.7.1 suggests, female dragons show a clear preference for males with red crests, speculate on why there are significant numbers of brown-crested and green-crested dragons in the Mythological Island population. How could you test your theory? (10%)

Results from previous captive-breeding experiments suggest that the three crest colours are associated with a single gene. From a genetic perspective, individual males can be homozygous green (green/green), or homozygous red (red/red), whilst heterozygotes (red/green) appear brown. Females have small crests that are always green irrespective of their genotype (green/green, red/green or red/red). You decide to investigate this further.

Whilst looking for short tandem repeats (STRs) that will allow you to design an appropriate breeding strategy, you notice that one of them seems to be co-inherited with the crest colour phenotype. You establish the pedigree of one particularly well-documented family of red-crested dragons. You make DNA from samples retained in Professor St George's freezer to enable you to determine the STR genotype of each individual. This allows you to generate the data shown in Fig. 3.7.6.

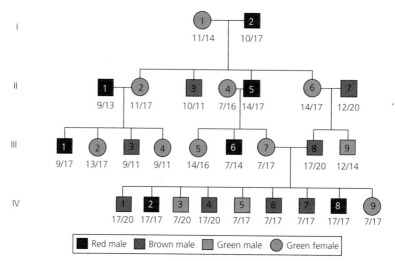

Figure 3.7.6   Pedigree of a typical captive-bred population of red-crested dragons. In addition to the crest colour, the genotype of a closely linked STR is shown for each individual.

Question 4: Redraw Fig. 3.7.6 showing the crest colour genotype of each individual. Where genotypes cannot be absolutely determined, explain why. In addition, indicate one position within the pedigree where there is evidence of recombination between the crest colour gene and the STR. (10%)

Question 5: You now have the approximate map position of the crest colour gene. Describe in outline a strategy that will enable you identify the DNA sequence of the crest colour gene. (5%)

Question 6: Speculate on what kind of protein the crest colour gene may encode. Explain how it may generate the three distinct crest colour phenotypes in males, but only one crest colour phenotype in females. (5%)

Professor St George is very impressed by your work and suggests you acquire samples of living crest tissue to help you investigate the genetic basis of crest colour in wild red-crested dragons. You are about to say something along the lines of 'But isn't that extremely dangerous?' when the Professor glances at her photograph of Sabra and says 'But please be very careful, as we don't want any more accidents.' You spend the rest of the afternoon learning how to use a dart gun that will allegedly temporarily anaesthetise the dragons.

Over the next few days you acquire several samples from adult dragons and bring them back to the laboratory for Next Gen RNA-seq analysis. You look carefully at the data, focussing on the crest colour gene you identified earlier. Next Gen RNA-seq reveals that all dragons with red crests have the RNA expression pattern shown in Fig. 3.7.7a. However, there were three different RNA expression patterns found in dragons with green crests, one of which was found in all green-crested dragons from the north of the island as shown in Fig. 3.7.7b. The green-crested dragons from the south of the island showed two distinct patterns of RNA expression as revealed in Fig. 3.7.7c and 3.7.7d.

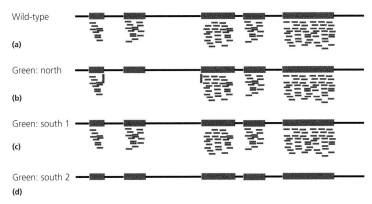

Figure 3.7.7   Next Gen RNA-seq of the crest-colour gene from adult red-crested dragons. The DNA sequence of the region as determined by the red-crested dragon genome project is shown by black line. The predicted five exon structure of the crest colour gene is shown as orange boxes within the genomic DNA sequence. Representative 100nt mRNA reads are shown in blue.

Question 7: Speculate on the nature of the mutations that cause green coloured crests in males from the north of the island (Fig. 3.7.7b) and the south of the island (Fig.s 7c & 7d)? (10%)

Question 8: What, if anything, does the Next Gen RNA-seq data tell you about the origin of the green-crested dragons in the north and south of Mythological Island? (5%)

Professor St George explains that one theory why the red-crested dragon population of Mythological Island is falling is that the dragons aren't breathing as much fire as they were 20 years ago. Indeed, it's clear that some adult red-crested dragons are unable to breathe fire at all. This characteristic was first noticed in dragons from the south, but now seems to be slowly spreading throughout the island.

Fire-breathing has two key functions for dragon well-being. It encourages mating and it enables the red-crested dragons to cook their food. As a consequence of their fire-breathing ability being compromised, red-crested dragons encounter two fundamental problems. Firstly, there are fewer matings and therefore fewer hatchlings. Secondly, because the immune system of adult red-crested dragons functions quite poorly, they are particularly susceptible to bacterial infection; so, if their food isn't cooked sufficiently, they are exposed to infection.

According to wikimania, the increase in the proportion of dragons that are unable to breathe fire may be linked to a catastrophic failure of the Mythological Island chilli pepper plants, which are thought to help fuel the fire-making ability of the dragons. Mythological chilli, *Capsicum incendia* (see Fig. 3.7.8), is thought to be the strongest chilli in the world, but a fungal infection that originated in the west of the island means it has almost disappeared from the entire island. However, Professor St George is highly sceptical of this theory.

Figure 3.7.8 Mythological Chilli (*Capsicum incendia*) the strongest chilli in the world.

Question 9: Why is Professor St George right to be sceptical of the theory that the decline of fire-breathing ability in red-crested dragons is due to the catastrophic failure of the Mythological chilli? (5%)

Previous studies of fire breathing in wild populations of red-crested dragons have allowed Professor St George and her colleagues to develop the Dragon-Activated Fire Test (DAFT). Essentially DAFT is a measure of the fire-breathing activity of dragons, which acts by measuring fire-induced damage on a 40-point scale, where 0 is no damage and 40 is torched.

It was your predecessor, Sabra de Silene, who had undertaken most of this work. DAFT activity levels are best estimated when the dragons are feeding, as this is when adult males and females both breathe fire. Sabra soon realised that a small subset of adult dragons from the south of Mythological Island produced little or no fire, and she coined the phrase 'extinguished' dragons. They score less than 5 on the DAFT scale. Sabra spent much of her short time on Mythological Island drugging and tagging red-crested dragons so she could develop a DNA test that would allow her to identify individuals and establish parentage within the wild population. It was this that led to the unfortunate incident from which she is now thankfully recovering.

Sabra estimated DAFT activity in fire-breathing dragons from the unaffected fire-breathing population in the north of the island, which she called her normal controls. She also estimated DAFTness in parents of the extinguished and found they were all fire breathers. Finally she looked at the siblings of the extinguished. Sabra's data is shown in Table 3.7.2.

Table 3.7.2   Numbers of red-crested dragons with specific DAFT activity.

	Units of DAFT activity			Total observed
	<5	11–25	26–40	
Normal controls	0	0	100	100
Parents of 'extinguished'	0	20	0	20
Siblings of 'extinguished'	4	11	5	20

Sadly Sabra was unable to analyse her results and Professor St George confesses she is confused. However, now that you have discredited the Mythological chilli decline theory, Professor St George suggests you investigate whether there is a genetic explanation.

Question 10: Propose a model for the genetic basis of the extinguished phenotype. (5%)

You focus your research on the extinguished dragons from the south of the island. In particular you wonder why all the extinguished dragons are only found in the south of the island. You look again at the data Sabra collected and start to look for any patterns. Suddenly it occurs to you that all the extinguished dragons have green crests! Although it should be stressed that not all dragons with green crests, even from the south of the island, have an extinguished phenotype. You tabulate the data, which is shown in Table 3.7.3.

Table 3.7.3   Number of individuals in the southern population of red-crested dragons with specific crest colour and DAFT activity phenotypes.

		Units of DAFT activity		
		<5	11–25	26–40
Crest colour	red	0	0	173
(males only)	brown	0	14	57
	green	23	6	40

However, you are well aware that there are two different green mutations in the crest colour gene within the red-crested dragon population in the south of Mythological Island, so you look again at the data but this time use Next Gen RNA-seq analysis to determine DAFT-activity for adult male red-crested dragons with each class of green crest mutation (see Table 3.7.4).

Table 3.7.4   Number of green crested individuals in the southern population of red-crested dragons with specific green mutations and DAFT activity phenotypes. The allele types, 1 and 2, refer to the green alleles described by the Next Gen RNA-seq data shown in Fig. 3.7.7. Genotypes 1/1 and 2/2 are homozygotes, whereas 1/2 is a green transheterozygote having one of each mutant allele (south 1 and south 2).

Green genotype	Units of DAFT activity		
	<5	11–25	26–40
Green: south 1/1	239	2	0
Green: south 2/2	0	0	407
Green: south 1/2	3	602	0

Question 11: What do the data in Table 3.7.3 and Table 3.7.4 tell you about the *extinguished* mutation? How will you use this data to help you find the *extinguished* gene? (10%)

Your data allows you to identify the precise sequence of the *extinguished* gene and notice that it is very close to a second gene called *infection-load-lethal* (*ill*). The protein encoded by the *ill* gene has DNA binding motifs and has previously been described as playing a key role in protecting red-crested dragons from bacterial infection. Figure 3.7.9 shows the chromosomal organization of the *extinguished* and *ill* genes, which clearly shows that they are transcribed in opposite directions.

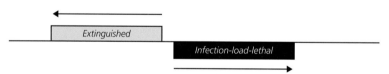

Figure 3.7.9   Chromosomal organisation of the *extinguished* and *infection-load-lethal* (*ill*) genes. The black line represents the DNA sequence with the positions of the coding regions of the two genes shown as coloured boxes. The arrows show that the *extinguished* gene is transcribed to the left and the *ill* gene is transcribed to the right.

You decide to look at mRNA expression levels of the *extinguished* and *ill* genes in gut tissues, as this is where the ability to breathe fire is focussed and this is also the part of the dragon that is most susceptible to bacterial infection. A summary of these data are shown in Table 3.7.5.

Table 3.7.5   Relative mRNA expression levels of the *extinguished* and *ill* genes in adult and juvenile males in wild-type and *extinguished* mutant red-crested dragons.

	Wild-type juvenile	Wild-type adult	Extinguished juvenile	Extinguished adult
extinguished **gene**	not detected	high	high	not detected
ill **gene**	high	not detected	not detected	high

Question 12: What are the normal expression patterns of the *extinguished* and *ill* genes? Speculate on why red-crested dragons may have evolved this pattern of expression. (5%)

Question 13: How are the expression patterns of the *extinguished* and *ill* genes altered in *extinguished* mutant juvenile and adult red-crested dragons? What consequences may these changes have for the phenotype of mutant red-crested dragons. (5%)

Question 14: Speculate on how a single mutation could cause the changes in *extinguished* and *ill* gene expression seen in the *extinguished* mutants. (5%)

Your 3-year research project has been a stunning success and you have made a number of significant discoveries:

- You have established the evolutionary relationship between red-crested dragons and the other eight extant species of dragon.
- You have identified three mutations in the crest colour gene, all of which lead to green crests
- You have identified and characterised the *extinguished* gene.
- You have identified and characterised the *ill* gene.

Professor St George is absolutely delighted by your work and is keen for you to continue but clearly you'll need to get some money to support you. She suggests that you put together a proposal for funding. She is particularly keen for the proposal to address the continuing problem of population decline in red-crested dragons. She thinks that your funding proposal is more likely to succeed if you use a series of different approaches.

Question 15: Design a series experiments that would allow you to investigate further the continuing problem of population decline in red-crested dragons. (10%)

Chapter 3.8

I Scream!

It all starts when your sister announces she's acquired a new boyfriend. That in itself doesn't surprised you, as he isn't her first 'new' boyfriend. Indeed, over the past couple of years quite a long list of new and entirely inappropriate boyfriends have been introduced, and an equally long list have suddenly disappeared, never to be seen again. You speculate that she is a one-woman 'black widow spider' and that somewhere nearby is a mass grave of the unwashed and unsuitable. However, this time it's different, apparently, and the boyfriend is coming to stay for the weekend.

The new boyfriend duly arrives, precisely on schedule. 'This is Juan,' your sister beams, 'Juan Cornetto,' in a way that makes him sound like an international man of mystery. Your first impression is one of complete and utter shock. Juan is very different from all previous incarnations of your sister's boyfriends, and not just because he seems to have grasped the concept of personal hygiene. He is immaculately dressed. He is also very cool. Yet, for some inexplicable reason, he has a huge crush on your sister.

As the weekend progresses, you learn all about the perfect boyfriend and his perfect family. The Cornettos live in a huge mansion near the southern Italian village of Gelato (see Fig. 3.8.1). Their family fortune apparently dates from the late 18th century when Great Grandad Cornetto developed a very sophisticated and very expensive ice cream that he successfully marketed to the Italian aristocracy.

Genetics? No Problem!, First Edition. Kevin O'Dell.
© 2017 John Wiley & Sons Ltd. Published 2017 by John Wiley & Sons Ltd.

Figure 3.8.1   The Cornetto family mansion on the hilltop near Gelato.

The Cornettos are ridiculously and almost embarrassingly rich and, as the eldest son of the eldest son of the eldest son in this traditional and patriarchal family, Juan will inherit everything. He is the rising star in the tightly knit Cornetto family. His recent suggestion that they diversify by making high-quality bespoke cakes has proved to be a masterstroke (see Fig. 3.8.2).

Figure 3.8.2   Juan's suggestion that the Cornettos diversify into making bespoke cakes had been hugely successful.

The weekend is a stunning success and your family spend much of the Sunday afternoon sitting in the garden, drinking the finest Italian wine that Juan has kindly brought with him from the Cornetto estate. Surely it can't get any better than this? In fact, during one fleeting moment when he isn't draped round your sister, Juan even seems to be interested that you've just finished a genetics degree and, funding permitting, are hoping to embark on a PhD.

You are just beginning to think that nothing could possibly go wrong, when out of the corner of your eye you spot your mother, who should never be trusted with a third glass of chilled Italian wine, stumbling in slow motion towards Juan and your sister (see Fig. 3.8.3). In the confusion your mother manages to pour the entire glass of wine over Juan's immaculately presented face. Your mother screams, but to your great surprise Juan simply sits there, expressionless and dribbling.

Figure 3.8.3   Your parents' garden moments before the unfortunate incident.

'Calm down' shouts your sister. 'It's the Cornetto curse. It's nothing to worry about. I just need to warm his lips up and he'll be OK.'

Having had his lips warmed in an entirely inappropriate fashion by your sister, Juan slowly regains his composure. He explains that when Great Grandad Cornetto first made the family's revolutionary ice cream, Portuguese rivals, the Sorvetes, put a curse on the Cornetto family, so they could never eat their own stunning tasty ice cream again. Ever since that fateful day, every time the cursed members of the Cornetto family try to eat their ice cream, their faces freeze and they start dribbling.

You don't believe a word of it and, in a quiet moment, persuade Juan to draw his family tree, starting with Great Grandad Cornetto. The pedigree, which shows how all the Cornettos are related to each other and which family members are affected by the Cornetto curse, is reproduced in Fig. 3.8.4. You soon begin to realise that Juan is not only a possible brother-in-law, but also a potential PhD opportunity.

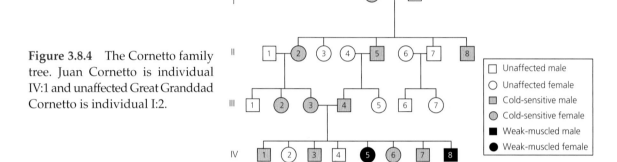

Figure 3.8.4   The Cornetto family tree. Juan Cornetto is individual IV:1 and unaffected Great Granddad Cornetto is individual I:2.

Question 1: Redraw the Cornetto family tree to reveal the evidence supporting the notion that the Cornetto curse is a genetically inherited condition. In particular speculate on why individuals IV:5 and IV:8 show muscle weakness, muscle atrophy and abnormal muscle morphology. (5%)

Question 2: Individuals III:3 and III:4 are married, but how else are they related and how does their additional relationship help you resolve the genetic basis of the Cornetto curse? (5%)

You suggest to Juan as politely as possible that his story about the Cornetto curse is complete and utter nonsense. Juan stares at you in disbelief, although for a moment or two you wonder whether he's been caught by a sudden gust of cold wind. You explain that as a young up-and-coming geneticist, your preliminary analysis of his family tree suggests there is strong evidence of an inherited genetic condition. Juan asks whether it would be possible for you to investigate this, but you point out that this kind of research is extraordinarily expensive and unfortunately as you don't have any PhD funding, there is very little you could do. 'We could fund you,' says Juan. You manage to feign total surprise.

A few days later you're flying first class to Italy to retrieve blood samples from a variety of Cornetto family members. You take the opportunity to read about other cases of reversible human cold-sensitivity of the type Juan exhibits, and discover they are often associated with mutations in sodium channel genes.

After a week in Italy, the time being necessary to ensure you retrieve replicate blood samples for each living member of the Cornetto family, you travel home. Meanwhile, money apparently being no object, Juan sends the blood samples to Professor Lynne Guiney, one of the worlds most renowned whole genome DNA sequencing specialists. A few days later Professor Guiney emails you the sequencing data and you set to work.

Question 3: How can an understanding of the Cornetto family pedigree and the whole genome DNA sequence of each family member help you identify the mutation that causes the reversible cold-sensitive paralysis? What problems might you encounter? (5%)

Question 4: How could you estimate the approximate date at which the mutation originally arose in the Cornetto family? (5%)

Now that your funding situation is secure, you visit your new PhD supervisor Doctor Nicholas Anyer, to discuss your ideas and progress. Doctor Anyer has a deserved reputation as being a brilliant but entirely unpredictable young researcher. He applauds your initiative, but even he is somewhat horrified that you've not followed the correct procedure in getting ethical approval for collecting DNA samples from the Cornetto family. You are about to point out that as a convicted serial shoplifter of Italian ready meals, he is hardly in a position to take the moral high ground, but you think better of it.

You discuss the sequence data with Dr Anyer, and identify a sodium channel gene, SCN4A, which has previously been associated with cold sensitivity, as the probable cause of the Cornetto phenotype. Doctor Anyer suggests that you compare the DNA sequences of the unaffected and cold-sensitive Cornetto SCN4A alleles, focusing on the sequence between codons 1271 and 1310. These sequences are shown in Fig. 3.8.5. It's only when he's left the building to do his weekly supermarket 'shopping', that you realise Dr Anyer has also provided you with a copy of the genetic code, which is reproduced here in Fig. 3.8.6.

```
1271
ATG/TAC/CTC/TAC/TTT/GTC/ATC/TTC/ATC/ATC/TTT/GGC/TCC/TTC/TTC/ACC/CTC/AAC/CTC/TTC/
     A                           T                                       T
Met/Tyr/Leu/Tyr/Phe/Val/Ile/Phe/Ile/Ile/Phe/Gly/Ser/Phe/Phe/Thr/Leu/Asn/Leu/Phe


1291
ATT/GGC/GTC/ATC/ATT/GAC/AAC/TTC/AAC/CAG/CAG/AAG/AAG/AAG/TTA/GGG/GGG/AAA/GAC/ATC/
                                                   C   A
Ile/Gly/Val/Ile/Ile/Asp/Asn/Phe/Asn/Gln/Gln/Lys/Lys/Lys/Leu/Gly/Gly/Lys/Asp/Ile
```

Figure 3.8.5 The normal Cornetto SCN4A sequence from codon 1271 to 1310 and the amino acids they encode. The Cornetto cold-sensitivity allele has the five DNA base changes shown.

Figure 3.8.6 The genetic code.

First letter	Second letter: T	Second letter: C	Second letter: A	Second letter: G	Third letter
T	TTT, TTC } Phe; TTA, TTG } Leu	TCT, TCC, TCA, TCG } Ser	TAT, TAC } Tyr; TAA Stop; TAG Stop	TGT, TGC } Cys; TGA Stop; TGG Trp	T C A G
C	CTT, CTC, CTA, CTG } Leu	CCT, CCC, CCA, CCG } Pro	CAT, CAC } His; CAA, CAG } Gln	CGT, CGC, CGA, CGG } Arg	T C A G
A	ATT, ATC, ATA } Ile; ATG Met	ACT, ACC, ACA, ACG } Thr	AAT, AAC } Asn; AAA, AAG } Lys	AGT, AGC } Ser; AGA, AGG } Arg	T C A G
G	GTT, GTC, GTA, GTG } Val	GCT, GCC, GCA, GCG } Ala	GAT, GAC } Asp; GAA, GAG } Glu	GGT, GGC, GGA, GGG } Gly	T C A G

Question 5: What is the consequence of each of the five DNA sequence changes in the cold-sensitive allele, and which may be responsible for the cold-sensitive phenotype in the Cornetto family? (5%)

For reasons that are never satisfactorily explained, Dr Anyer is unable to return to work in the laboratory for a few weeks, so you take the opportunity to discuss your DNA sequencing results with Doctor Marc Aronie. He is an expert on *Drosophila* neurogenetics and on hearing about Dr Anyer's unscheduled absence, he suggests that you model Cornetto cold sensitivity in the fruit fly *Drosophila melanogaster*. You spend the next few days in his laboratory, shown in Fig. 3.8.7, investigating how you can use *Drosophila* as a model of human genetic disease.

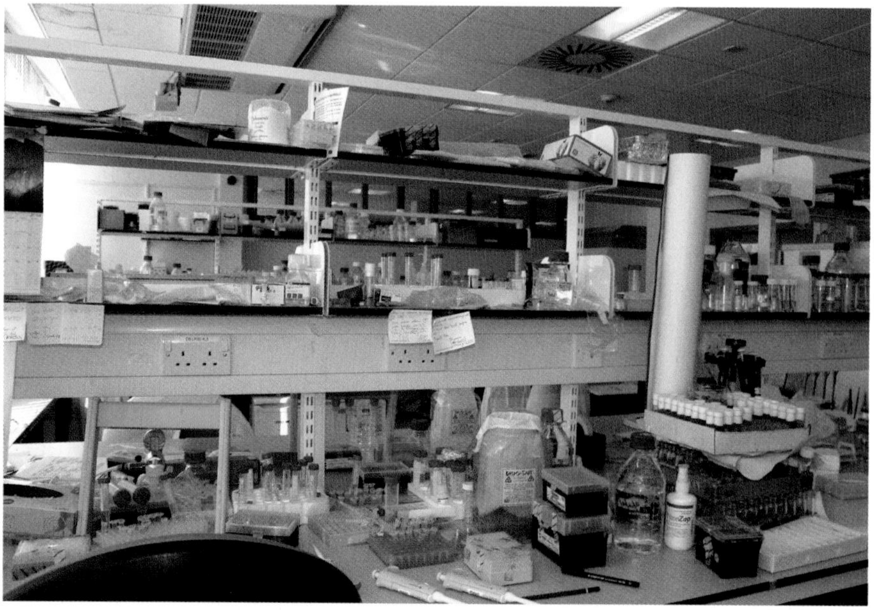

Figure 3.8.7 Doctor Marc Aronie's fly laboratory.

You discover the following relevant facts:

- *Drosophila* have a single sodium channel gene called *paralytic* (*para*, see Fig. 3.8.8).
- The *paralytic* gene undergoes alternative splicing and RNA editing.
- Null mutations of the *paralytic* gene are recessive lethal.
- Researchers use the *P*-element transposon to generate transgenic flies.

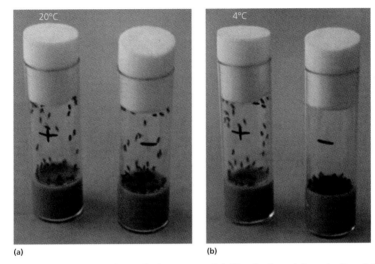

(a)   (b)

Figure 3.8.8   Wild-type (+) and *paralytic* mutant (−) flies before (a) and after (b) cold shock.

A stylized image of a *P*-element used to create transgenic flies is shown in Fig. 3.8.9.

Figure 3.8.9   Stylized image of the *P*-element used for creating transgenic *Drosophila* in this study, showing the 31 bp inverted repeats that define the ends of the *P*-element, the approximate position of the *Gene of Interest* (in this case the *paralytic* sodium channel gene) and a wild-type copy of the *white* eye colour gene of *Drosophila*.

Question 6: Why does the *P*-element transgene contain: (a) the 31 bp inverted repeats; and (b) the wild-type copy of the *white* eye colour gene? (5%)

Question 7: What transgenic flies are you going to make to identify the DNA sequence change (mutation) in the human SCN4A gene that causes the Cornetto cold-sensitivity phenotype? Specifically, what DNA sequence(s) should the *Gene of Interest* in the *P*-element (see Fig. 3.8.9) actually have? What problems might you encounter and how might you resolve these? (5%)

Question 8: When you have made your *P*-element transgene you will need to inject it into the germ line of a host fly. That host fly could be wild-type, heterozygous or homozygous mutant for the *paralytic* gene, but what would be the best *paralytic* genotype and why? (5%)

You generate a series of transgenic flies that are homozygous wild-type for the endogenous *paralytic* gene of *Drosophila* and that also carry a single copy of one of the *paralytic* transgenes you've created. Among these new transgenic lines are four strains that carry independent insertions of the I1273 *paralytic* transgene (called I1273 (1) to I1273 (4)) and another four that carry independent insertions of the R1306 *paralytic* transgene (called R1306 (1) to R1306 (4)).

You develop a cold-exposure test that involves using one of the Cornetto family's glass-fronted ice-cream cabinets. You set the temperature control at 4°C and place tubes of flies inside. This allows you to watch the behaviour of the flies under cold conditions. Flies that are paralysed by the cold fall over, so you simply count the number of flies that remain upright every 20 seconds over a 300-second observation period. The cold sensitivity of a wild-type strain and the eight transgenic lines described earlier are shown in Fig. 3.8.10.

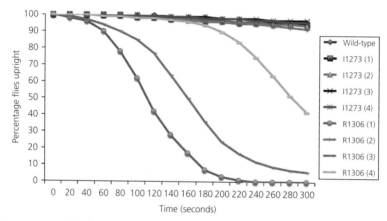

Figure 3.8.10   Percentage of wild-type and transgenic flies (I1273(1–4) and R1306(1–4)) remaining upright when exposed to a cold temperature of 4°C.

Question 9: What does the cold-exposure assay reveal about the genetic basis of cold sensitivity? (5%)

Question 10: Even though transgenic lines R1306 (1) to R1306 (4) have a single copy of the same transgene, they show marked differences in their sensitivity to cold. Speculate on why this might be and design an experiment that would test your theory. (10%)

Your work on the *Drosophila* model of Cornetto cold sensitivity is a great success, but with the continued absence of Dr Anyer, and no definitive news of when he may return, you realise that if you are ever going to develop a treatment or cure for your sister's boyfriend, you are going to need to create a mouse model of Cornetto cold sensitivity.

However, you initially struggle with deciding how to take the project forward. Your data strongly suggests that the Cornetto cold sensitivity is caused by the G1306R mutation in the human SCN4A gene. However, like humans, mice also have nine sodium channel genes, so how are you going to decide which mouse sodium channel gene that when mutant would cause the cold-sensitive phenotype? You decide to contact the renowned evolutionary biologist Doctor

Ravi O'Leigh (a man of complex Irish/Indian heritage and who would be an interesting evolutionary research project in his own right). Doctor O'Leigh kindly sends you some data on the evolutionary relationship of the nine human sodium channel genes. This image is reproduced in Fig. 3.8.11.

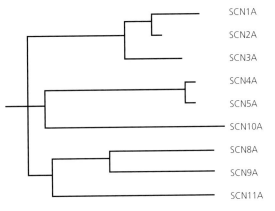

Figure 3.8.11 Apparent evolutionary relationship between the nine human sodium channel genes, SCN1A to SCN5A and SCN8A to SCN11A.

Unfortunately, Dr O'Leigh hasn't sent you any details about how he derived the evolutionary tree shown in Fig. 3.8.11, so you have to work this out for yourself.

Question 11: How is an evolutionary tree such as that presented in Fig. 3.8.11 constructed? How can you be confident that its shape is reliable and robust? (5%)

Question 12: Like humans, mice also have nine sodium channel genes. How could you determine which, if any, of the mouse genes is equivalent to the human SCN4A gene? What problems might you encounter? (5%)

Your work with Dr O'Leigh clearly shows that mice do have a sodium channel gene that sequence data suggests is equivalent to human SCN4A. You sensibly decide to generate a mouse model of the Cornetto cold sensitivity by creating a transgenic mouse with the SCN4A G1306R missense mutation.

The transgenic mouse expert in the research institute where you work is Professor Penny Rigartay. She is delighted to have the opportunity to help you, especially as the Cornetto Foundation so generously funds your work. As the cold-sensitivity mutation is dominant in both humans and flies, she suggests you create a 'knock-in' transgenic mouse that has a modified R1306 version of its host SCN4A gene.

Question 13: Design a construct that would allow you to make mouse with a 'knock-in' SCN4A R1306 mutation. Briefly describe the strategy that will allow you to generate a 'knock-in' transgenic mouse with an SCN4A R1306 mutation. (10%)

You arrange to discuss your transgenic mouse plans with Professor Rigartay, but when you enter her office you are surprised to see Juan sitting there, looking every bit like the perfect young entrepreneur he aspires to be. 'Juan has a proposal for you,' says a very smiley Professor

Rigartay. You have a moment of panic as your first thought is that Juan should be directing any proposals to your sister, but then you realise it must be something to do with your research. 'It's about your mouse model,' Juan says reassuringly. 'If it works, then the Cornetto Foundation want to use it to develop a new research facility in our home town of Gelato,' he continues. 'We want to call our facility *Gen-Italia* and were rather hoping that when you finish your PhD you'd like to join us.'

Everything is a bit of a blur, you're not sure what you want your future plans to be, but it seems too good an opportunity to miss. 'I'd be delighted to,' you say, trying and failing not to seem too keen. 'On condition that you don't ask me to draw the new company logo,' you add.

The mouse model work takes up a lot of your time and it's 18 months since you worked with the fly stocks you created. One afternoon you have the opportunity to visit your research institute's curator of fly stocks, Mr Rick Cotter. He's been looking after the transgenic fly strains you created at the beginning of your research project. Rick has a bit of a reputation as doing the minimum amount of work necessary to retain his job. In 30 years has never been known to use his initiative, so you feel it's wise to recheck the stocks he's been looking after.

When you sequence each of the four transgenic strains that originally possessed a single insertion of the R1306 transgene, you're reassured that that all four transgenic lines retain the R1306 transgene. In each case the DNA sequence of the R1306 transgene is complete and unaltered. You also check the cold-sensitivity phenotypes of the four strains, and this data (along with that of the cold-sensitivity data from the same four strains when they were tested 18 months ago) is presented in Fig. 3.8.12.

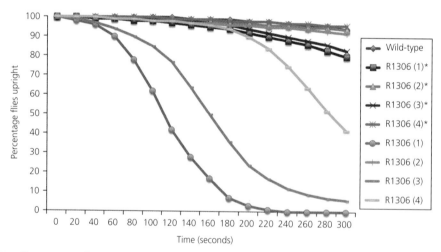

Figure 3.8.12   Percentage flies remaining upright when exposed to a cold temperature of 4°C. The data from the wild-type and transgenic strains R1306 (1–4) are identical to those presented in Fig. 3.8.10. Transgenic strains R1306 (1*–4*) are the same strains as R1306 (1–4) but have been tested 18 months later.

At first you are surprised that the R1306 transgenic strains have lost their cold-sensitivity over the past eighteen months. Rick assures you that he kept the strains in precisely the way you asked, in that all strains were kept as homozygous for the endogenous wild-type *paralytic* gene, and homozygous for the R1306 *paralytic* transgene.

Question 14: Speculate on why the cold sensitivity of the four original transgenic strains R1306 (1–4) has in some cases apparently changed over the 18 months since they were last tested. (10%)

It is at this point that Dr Anyer reappears, and you're delighted to have a chance to discuss the strange results for fly cold sensitivity with him. He is really excited about the data and asks you to explain precisely how you originally made the fly strains, whether you kept any molecular samples from those original strains, and how confident you are that the coding sequences of the endogenous G1306 *paralytic* gene and the R1306 *paralytic* transgene in the existing cold-tolerant strains are absolutely identical to the strains you originally created 18 months ago. You check everything and establish the following facts:

- You have DNA samples from each of the four *paralytic* transgenic lines (R1306 (1–4)) from 18 months ago stored in your freezer.
- You have DNA samples from each of the four *paralytic* transgenic lines (R1306 (1*–4*)) that are alive today and are the descendants of the four *paralytic* transgenic lines (R1306 (1–4)).
- When you compare the coding sequences of the G1306 endogenous *paralytic* gene and the R1306 *paralytic* transgene from the 18-month-old DNA samples (R1306 (1–4)) to those from the existing stocks alive today (R1306 (1*–4*)), you find that the SCN4A coding sequences are identical. So the change in cold sensitivity is not associated with new mutations within the coding region of the endogenous *paralytic* gene, nor is it within the *paralytic* transgene.
- Over the same 18-month period, the R1306 transgenes have not moved from their original insertion sites.
  You have a few months left at the institute before you leave for *Gen-Italia* in Gelato and you want to use your time as productively as possible.

Question 15: Design an experiment that would allow you to identify whether the loss of cold sensitivity in the transgenic lines has a genetic or environmental cause. (5%)

You discover that the loss of cold sensitivity in the transgenic lines is genetic.

Question 16: Design an experiment that would allow you to identify the genetic cause of the loss of cold sensitivity in the transgenic lines. (10%)

Chapter 3.9

The Nuns of Gaborone

It's only as your flight touches down in Sir Seretse Khama International Airport that you begin to relax. You can hardly believe it's happening! You've never been to Botswana before, and now it's going to be your home for the next 3 years. Your plan is to study the blue wildebeest (*Connochaetes taurinus*) of Botswana's Moremi Game Reserve (see Fig. 3.9.1). You've heard that these magnificent animals are still found in large numbers across much of Botswana and its neighbouring countries, so they may be the ideal large mammal for studying the definition of a species. Do all blue wildebeest really belong to a single species, or can they be subdivided into distinct population or species groups? You're hoping that your background as a recently graduated genetics student will help you answer this question.

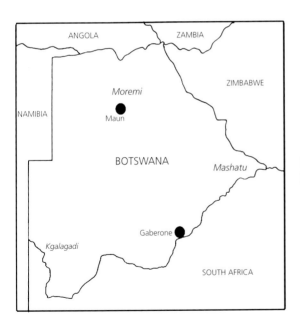

Figure 3.9.1 Map of Botswana, showing Gaborone, Maun and the Moremi Wildlife Reserve.

Genetics? No Problem!, First Edition. Kevin O'Dell.
© 2017 John Wiley & Sons Ltd. Published 2017 by John Wiley & Sons Ltd.

You phone your father once you're safely inside the terminal building and reunited with your bags. Sadly, you never had the opportunity to meet your mother, and it's not something your father seems to want to talk about, but your father has always been there for you and over the years he has been your greatest supporter. He is absolutely delighted that you're spending some time in Botswana. He lived in Gaborone for a while when he was younger and he still has some contacts there. 'And remember,' he was fond of saying, 'if anything goes wrong, contact the No. 1 Ladies' Detective Agency!'

You spend a few days in Gaborone before you travel onto the Moremi Game Reserve, and at your father's insistence stay in a hostel owned by the Nuns of Gaborone, which is where your father stayed many years ago. He has very fond memories of his time in Gaborone and as he's now a successful entrepreneur, he is delighted to have the opportunity to support the Nuns financially. The Nuns of Gaborone are very pleased to see you.

A week later, you fly to Maun Airport, before travelling on to the Moremi Game Reserve that covers much of the eastern side of the Okavango Delta. It is home to an extraordinary variety of dryland and wetland animals and is a wonderful place to explore. Chief's Island and the Moremi Tongue, which incorporates Mboma Island, are the main areas of dryland in this wetland reserve. You are interested in whether local populations of blue wildebeest (see Fig. 3.9.2) have adapted to the distinct wet and dry environments within the Moremi Game Reserve and whether their huge herds have formed distinct species groups (see Fig. 3.9.3).

Figure 3.9.2   An adult wildebeest.

Figure 3.9.3   A huge herd of wildebeest.

Question 1: Define the term 'species' and, in particular, explain how analyses of DNA sequences can be used to determine whether two populations should be regarded as one species or two. (5%)

Your accommodation at the Moremi Game Reserve is within a complex again run by the Nuns of Gaborone. In the part of the complex where you're staying, there are several game reserve workers of European descent who seem to have stepped through a time warp from the colonial era. The Moremi Chapter of the Nuns of Gaborone live in the other half of the complex.

It's only now that you discover all is not well with the blue wildebeest populations of Moremi. In fact, for inexplicable reasons, this year has seen a catastrophic decline in the blue wildebeest population across Botswana and its neighbouring countries. Local researchers suspect that the decline is linked to a drought that is gripping the country, and have collected data regarding changes in blue wildebeest population sizes across Botswana. But how can they confirm their theory? You see an early opportunity to impress and you contact the Botswana University Meteorological Station to ask them to supply countrywide rainfall data. Figure 3.9.4 shows your plot of rainfall against blue wildebeest decline in 20 populations across Botswana.

Figure 3.9.4   The relationship between rainfall and estimated blue wildebeest population levels in a variety of areas within Botswana. Each blue dot represents a different population of blue wildebeest within Botswana. The percentage population change is given as this year's population size as a percentage of last year's population size. The percentage of average rainfall indicates the rainfall level for the current year as a proportion of the mean rainfall for the previous 10 years.

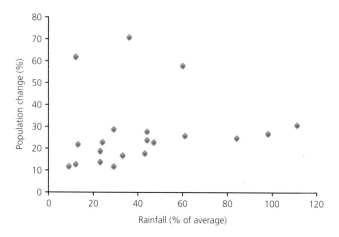

To determine whether there is a relationship between rainfall and population decline, you determine the correlation coefficient (r), which you calculate as 0.13. For 18 degrees of freedom, the cut-off for $p = 0.05$ is 0.44, which suggests there is no relationship between rainfall and population decline.

You notice that 3 of the 20 populations studied (your host population of Moremi in the north, Kgalagadi in the south-west and Mashatu on the Limpopo River in the east) seem to have much higher survival rates than the rest of the populations.

Question 2: Is the decline in blue wildebeest populations primarily associated with the drop in rainfall, or is there likely to be another explanation? Explain. (5%)

You decide to study the Moremi, Kgalagadi and Mashatu populations further.

If the decline in blue wildebeest populations in Botswana is not a consequence of the drought, then what is the cause? You spend several days reading the blue wildebeest research literature, and it soon becomes apparent that the biggest problem for blue wildebeest is their interaction with people. Direct hunting isn't a particularly big problem; however, fences, roads and railways all create barriers on the established blue wildebeest migration routes. Indeed, fences constructed to restrict the movement of domesticated cattle herds are a particular problem. You wonder whether it's competition from cattle that is causing the decline in blue wildebeest numbers.

You discover some old research data from 1920s Bechuanaland (the colonial era name for Botswana) discussing a viral infection that had devastated the cattle population. The survivors of that population, who must have inherited a natural immunity to the infection, are the ancestors of the current cattle population of modern-day Botswana. You also read that viral infections can occasionally pass between species by a process called xenoinfection. In the 1920s, the diseased cattle suffered from severe immune dysfunction and as a consequence were highly prone to further infections. Interestingly, they also exhibited a severe weight-loss phenotype. You notice that blue wildebeest from the populations whose decline had originally been attributed to the drought, also exhibited significant weight loss. You decide to investigate whether the current decline in blue wildebeest populations is caused by the same virus that devastated the cattle populations in the 1920s.

Question 3: Explain why a virus that is endemic in a cattle population where it causes few health issues, might be potentially lethal when it crosses the species barrier and infects blue wildebeest? (5%)

Question 4: How could you determine that an infective agent such as a virus is linked to the recent decline in blue wildebeest population size? What problems might you encounter? (5%)

Question 5: How could you determine whether blue wildebeest can be infected asymptomatically? (5%)

Question 6: How could you determine whether the blue wildebeest were infected with the same virus that can infect cattle? (5%)

Your theory seems to be correct. The population decline in blue wildebeest seems to be associated with a virus that they've recently acquired from domestic cattle. You're delighted with your progress and return to your room in the complex for a well-earned rest. Your joy is short-lived, however, as you're horrified to discover the door to your room is wide open and your iPod, a going-away present from your aunt, is missing.

You're curious to discover whether the apparent lack of population decline in the Moremi, Kgalagadi and Mashatu blue wildebeest is actually the result of an inherited genetic resistance to the virus; however, there is an obvious problem. To study inheritance patterns you'll need some data from breeding experiments, and blue wildebeest take about a year to mature. So, even if you start these experiments now, it'll be at least a year before you have any data. You ask your colleagues whether there are any local research groups who might already have recorded the breeding patterns of blue wildebeest and are prepared to share their data. They promise to ask around.

A few days later you receive a surreal phone call from a man who introduces himself as Rra Gnou. 'I am a member of WOBBLE,' he says. 'We run the *Wildebeest of Botswana Biodiversity and Local Environment* project, and would like to help you with your research,' he continues. Astonishingly, just 3 days later you have access to all the mating data you could possible need! As WOBBLE have already mated blue wildebeest from different populations, all you have to do is infect the animals to determine whether they are resistant or susceptible. You establish the following facts:

- As determined earlier, there seem to be three populations of resistant blue wildebeest: Moremi, Kgalagadi and Mashatu.
- Crosses between resistant blue wildebeest within a population (Moremi × Moremi, Kgalagadi × Kgalagadi or Mashatu × Mashatu) always generate resistant blue wildebeest.
- Crosses between resistant and susceptible blue wildebeest within a population (Moremi × Moremi or Kgalagadi × Kgalagadi or Mashatu × Mashatu) sometimes gives a mix of resistant and susceptible offspring, and sometimes gives only susceptible offspring.
- Crosses between susceptible blue wildebeest within populations (Moremi × Moremi or Kgalagadi × Kgalagadi or Mashatu × Mashatu) usually give susceptible offspring, but occasionally generates resistant offspring.

Question 7: What do the data based on crosses between blue wildebeest within populations suggest about the genetic basis of resistance within the three populations? (5%)

Further analysis reveals that crosses between populations (Moremi × Kgalagadi or Moremi × Mashatu or Kgalagadi × Mashatu) gives precisely the same data as crosses within a population described earlier.

Question 8: What do the data based on crosses between blue wildebeest of the different populations suggest about the genetic basis of resistance in Botswana blue wildebeest? (5%)

As you've already identified the virus causing population decline in blue wildebeest, you focus your studies on a gene encoding a wildebeest chemokine receptor. The virus is believed to use this receptor to enter host T cells in cattle. The virus then kills these cells and it is this that leads to the progression and development of the disease. You collect tissue samples from sensitive and resistant blue wildebeest from each of the three identified populations. This allows you to undertake a whole genome RNA-seq analysis. The RNA-seq data for the chemokine receptor is shown in Fig. 3.9.5.

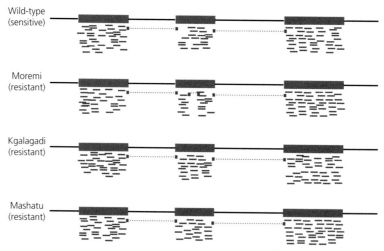

Figure 3.9.5   RNA-seq data from the chemokine receptor in the wild-type (sensitive) population, as well as from survivors from each of the three apparently resistant populations: Moremi, Kgalagadi and Mashatu. A black line shows the DNA sequence of the region as determined by the blue wildebeest genome project. The predicted three exon structure of the chemokine receptor gene is shown as orange boxes within the genomic DNA sequence. Representative 100 nt mRNA reads are shown in blue.

Question 9: What does the RNA-seq data reveal about the possible genetic basis of resistance in the Moremi, Kgalagadi and Mashatu populations of blue wildebeest. (10%)

You are delighted that you have discovered some provisional evidence supporting your theory of a viral resistance gene in blue wildebeest. You return to the complex to celebrate your discovery. However, your excitement is short-lived as you discover the new lock on the door to your room has been forced open and your watch, which was a leaving present from your father, is missing.

Meanwhile, your collaborators at WOBBLE send you the DNA sequence data from the chemokine receptor gene in susceptible and resistant blue wildebeest from the three populations. A summary of the WOBBLE codon data is shown in Table 3.9.1. It shows DNA sequence differences at polymorphic codons in the six groups studied. The genetic code is shown in Fig. 3.9.6.

Table 3.9.1 DNA sequence differences at ten codons in the chemokine receptor gene in susceptible and resistant blue wildebeest from three populations, Moremi, Kgalagadi and Mashatu. Apart from the changes listed here, which were also revealed by the RNA-seq analysis described in Fig. 3.9.5, no other DNA/RNA sequence differences were found.

Codon	Moremi		Kgalagadi		Mashatu	
	Susceptible	Resistant	Susceptible	Resistant	Susceptible	Resistant
7	CCT	CCT	CCC	CCC	CCC	CCC
33	TAC	TAC	TAC	TGC	TAC	TAC
34	CCC	CCC	CCC	CCC	CCA	CCA
73	CTA	ATA	CTA	CTA	TTA	TTA
134	AGT	AGT	AGC	AGC	AGT	AGT
147	TGT	TGT	TGT	TGT	TGT	CGT
201	ATA	ATA	TTA	TTA	ATA	ATA
222	CAT	CAT	CAC	CAC	CAC	CAC
334	AGG	AGA	AGG	AGG	CGG	CGG
336	TCG	TCG	TCT	ACT	TCT	TCT

Figure 3.9.6 The Genetic Code.

Question 10: What does the DNA sequence data suggest might be the genetic basis of viral resistance in the Moremi, Kgalagadi and Mashatu populations of blue wildebeest? How would you test your theory and what alternative explanations might there be? (10%)

Question 11: What does the DNA sequence data suggest about the evolutionary origin of resistance in Botswana's blue wildebeest? (5%)

Question 12: Suggest two possible mechanisms by which a blue wildebeest heterozygous for a chemokine receptor mutation may have partial resistance to viral infection. (5%)

Interestingly, you have noted that there are always survivors in blue wildebeest populations even where there are no chemokine receptor gene mutations. Further analysis of your sequencing data reveals that the gene encoding a chemokine ligand (which can reduce the amount of chemokine receptor on the surface of T cells) shows gene copy number variation. You calculate ligand gene copy number in 100 infected and 100 uninfected blue wildebeest in which the chemokine receptor mutation is absent, to determine whether there is a relationship between ligand gene copy number and infection. All the animals tested live in an area where infection is rife, so it is very likely that all the animals were exposed to the virus. These data are shown in Fig. 3.9.7 and Table 3.9.2.

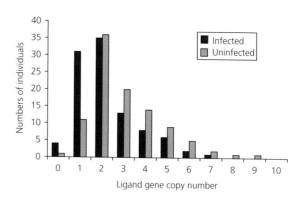

Figure 3.9.7   Number of infected and uninfected blue wildebeest with a specific copy number of the chemokine ligand gene. $N = 100$. The raw data are shown in Table 3.9.2.

Table 3.9.2   Number of infected and uninfected blue wildebeest with a specific copy number of the chemokine ligand gene. N=100. These data are taken from Fig. 3.9.7.

	Ligand gene copy number										
	0	1	2	3	4	5	6	7	8	9	10
infected	4	31	35	13	8	6	2	1			
uninfected	1	11	36	20	14	9	5	2	1	1	

Question 13: Calculate the mean ligand gene copy number in infected and uninfected individuals. Do these data suggest there is any affect of ligand gene copy number on blue wildebeest survival rates when they are exposed to the virus? (5%)

You return to your room in the complex and this time find the window wide open. Your painting of Charles Darwin by the famous artist Andy Warthog, shown in Fig. 3.9.8, is missing.

Figure 3.9.8   Painting of Charles Darwin by the renowned artist Andy Warthog.

Your 3-year placement in Botswana is coming to an end, and whilst you're pleased with the progress you've made in determining the cause of population decline in blue wildebeest, you wish there was something more practical you could do. It's then that your new friends from WOBBLE tell you about their new research funding stream that is targeted at investigations into threatened African wildlife. They invite you submit an application and ask you to write a brief summary of the key findings of your research. You include an image of the virus that you think is the cause of the infection, which is shown in Fig. 3.9.9.

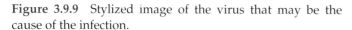

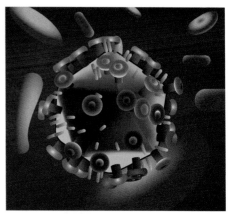

Figure 3.9.9 Stylized image of the virus that may be the cause of the infection.

Question 14: For your application to WOBBLE, propose a model showing how variation at the chemokine receptor gene and chemokine ligand gene might affect survival rates when blue wildebeest are exposed to the virus. (10%)

Question 15: For your application to WOBBLE, propose in outline how you might develop one treatment for infected blue wildebeest, and one preventative measure that may help blue wildebeest survive the viral infection. (5%)

You're pleased with the proposal you've written but want to have a final read through before you submit it. You nip out of your room to collect a gnuburger from the kitchens and, though you're only away for a few minutes, when you return to your room, your laptop is missing!

It's been a hugely successful 3 years and you're excited about having the opportunity to develop treatments and cures for the blue wildebeest of Botswana. However, there is one outstanding issue. Why do things continually go missing from your room and, even though you find it hard to imagine, is there a thief living in the complex? You could follow your father's advice and contact the No. 1 Ladies' Detective Agency, but you're confident that you have the technology to resolve this yourself. You set a trap! That evening at dinner, you make a point of telling everyone that you're going to spend the night observing nocturnal behaviour in blue wildebeest. However, while you are out, you leave the door to your room unlocked with your replacement iPod clearly in view. Somewhat cunningly, you work on the edges of the iPod so they are razor sharp and you are not at all surprised when you return to your room the following morning and find that your iPod has been replaced by a moderate quantity of blood.

You spend the next couple of days in the laboratory establishing the short tandem repeat (STR) profile of the blood sample and compare it to the Moremi Game Reserve online STR database that holds the profiles of the entire workforce. This database is important as it may be necessary to identify human remains if there is an unfortunate incident with a hungry lion. The STR profile data is shown in Table 3.9.3.

Table 3.9.3 STR profiles of twelve people and the blood sample from your room at 13 unlinked STRs. All 12 people live in the Nuns of Gaborone complex at Moremi Game Reserve. Your age and sex is unknown to the author at the time of publication.

Character	Age	Sex	STR1	STR2	STR3	STR4	STR5	STR6	STR7	STR8	STR9	STR10	STR11	STR12	STR13
Blood Sample		female	25/26	20/24	42/48	35/39	51/54	29/30	48/49	23/29	65/66	81/82	15/19	56/58	14/14
Captain Keith	67	male	20/26	15/16	41/48	31/34	49/50	23/31	46/52	22/22	61/68	86/88	13/20	60/63	14/19
Corporal John	72	male	19/23	18/23	41/46	32/41	59/59	25/25	44/48	30/31	65/65	82/86	13/14	68/70	14/19
Major Roy	39	male	22/28	18/21	43/43	38/39	52/58	23/28	49/50	23/24	64/70	87/88	13/21	61/70	14/19
Sister Maria	23	female	21/24	16/23	42/45	36/38	52/57	24/25	45/47	25/30	62/64	85/86	15/16	59/66	14/14
Sister Andrea	46	female	24/25	15/21	46/49	38/40	54/59	26/31	45/51	26/28	65/67	84/87	20/20	57/62	14/14
Sister Anna	72	female	25/26	20/24	42/48	35/39	51/54	29/30	48/49	23/29	65/66	81/82	15/19	56/58	14/14
Sister Boitumelo	31	female	20/23	17/18	44/47	33/33	56/58	28/30	44/47	26/27	61/66	80/84	14/18	62/71	14/14
Sister Gorata	56	female	23/27	15/24	43/49	36/41	52/53	27/27	45/48	22/27	68/69	81/89	21/21	61/72	14/14
Sister Kagiso	31	female	20/23	17/18	44/47	33/33	56/58	28/30	44/47	26/27	61/66	80/84	14/18	62/71	14/14
Sister Tapiwa	21	female	26/28	23/24	44/45	33/41	53/57	27/31	50/50	28/29	63/68	79/87	14/17	69/73	14/14
Spyros	25	male	21/24	19/23	43/45	34/37	57/59	23/29	45/50	25/30	62/64	80/81	16/17	65/66	14/19
You	**	**	22/25	19/21	46/47	32/40	52/59	26/27	46/51	24/28	63/67	79/84	19/20	57/59	14/**

Question 16: What does the data indicate regarding the perpetrator of the crime? (5%)

Question 17: Several of the other people living in the hostel are related. Who are the twins, who are the siblings and who are the parent and child? And what are you going to tell your father? (5%)

Chapter 3.10

Poissons Sans Yeux

It was perhaps one of the greatest triumphs of your childhood. It was the day your father couldn't answer your question! You must have been about 6 or 7 years' old at the time. Your father was a very keen tropical fish enthusiast and you remember him spending hours feeding, cleaning and generally fussing over the little aquatic animals. You'd always enjoyed following your father on his frequent tours of local pet shops and were keen to impress him with your growing knowledge of all things fishy. Looking back, you realise that he must have had a remarkable amount of patience with this small person who fired off an endless barrage of questions. Your father always found the time to reply and you gradually gained a substantial understanding of the world of tropical fish.

Then came that magic day. It was the opening of Sue Narmy's Tropical Fish Emporium in the city where you lived. You entered the new shop and set about searching for a new addition to your shoal, when you noticed some strange almost colourless fish (see Fig. 3.10.1). Miss Narmy explained that they were a newly discovered species from the West Henriette cave system in Louisiana and that they were completely blind, having a strange darkly pigmented spot where their eyes would have been. Your first fairly unimaginative question was something like 'How do they know which direction to swim in?' Then you remembered a story your father had told you that fish were like people because they behaved differently at different times of the day as they could sense light and dark cues from their environment. 'How do they tell the time?' you asked, and for once your question was met by silence.

Genetics? No Problem!, First Edition. Kevin O'Dell.
© 2017 John Wiley & Sons Ltd. Published 2017 by John Wiley & Sons Ltd.

Figure 3.10.1   Fish from the West Henriette cave system in Louisiana living in a tank at Sue Narmy's Tropical Fish Emporium.

Now, many years later, you are about to start a PhD in the laboratory of renowned researcher Professor Millie Second. Fortunately, she has just received funding from the entrepreneur Sir K. D. N. Writham to investigate whether cave-dwelling animals live in a 24-hour world. Your childhood story, coupled with your recent genetics degree, has convinced her that you are the person to initiate the project.

On your first day, you immerse yourself in the very limited research literature on the fish from the West Henriette cave system in Louisiana and you are astonished that they are only found in this one geographical location. It seems that the fish were first discovered in 1755 by the little-known French explorer Pierre le Pecher, who named the cave system after Louise Henriette de Bourbon, Duchess of Orléans. Rather unimaginatively, Monsieur Pecher named the fish *Poissons sans yeux chez Henriette Ouest* (eyeless fish from West Henriette). As the region became more anglicised, the English-speaking settlers, perhaps encouraged by observing the fish frequently but gently bumping into each other, began to refer to the species by its French acronym, 'psycho'. Professor Second has received special permission from the Louisiana legislature to use these rare psychofish in her research.

After discussions with Professor Second, you agree to start by investigating whether or not psychofish exhibit a 24-hour activity rhythm. According to the published scientific literature, fish are generally more active during their perceived daytime than at their perceived nighttime. As there are relatively few psychofish and they are very valuable, you appreciate that any experiment you carry out should in no way put their health and their lives at risk.

You initially think this is going to be fairly easy because, surely, all you have to do is watch the psychofish to discover when they're active. However, when Professor Second introduces you to the fish population room shown in Fig. 3.10.2, you discover that there are over 100 tanks containing psychofish. She tells you that each tank contains a population of psychofish derived from a single wild-caught pregnant female; these stocks are irreplaceable. Clearly, simply watching 100 tanks of fish to see how they behave is not a realistic experimental strategy!

Figure 3.10.2 Professor Millie Second's fish population room.

Question 1: Design a non-invasive experiment that would allow you to determine whether the psychofish populations in each tank live in a 24-hour world. (5%)

Your initial data suggest that psychofish do indeed show rhythmic behaviour. However, as the data shown in Fig. 3.10.3 reveal, the activity rhythm of the laboratory-bred psychofish is only around 22-hours long.

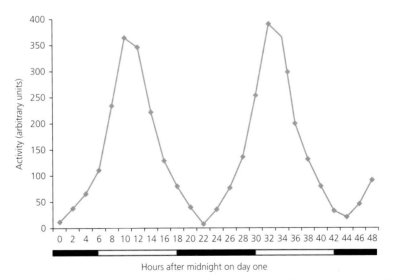

Figure 3.10.3 Activity rhythm of psychofish over 48 hours under otherwise standard laboratory conditions in the dark. Time 0 is midnight on day one, the start of the study. Note that before the experiment the fish were living in a strict 12-h light:12-h dark environment. The bar under the figure indicates when the lights would have been on (white) or off (black) if the 12:12 light:dark regime had been continued.

You are curious to know why the psychofish show an activity rhythm of 22-hours rather than the traditional 24-hours, as you've been careful to replicate the natural conditions of the West Henriette cave system as much as possible. However, you realise that a dark cave may be colder than Professor Second's laboratory, so you repeat your experiment at different temperatures. These data are shown in Fig. 3.10.4.

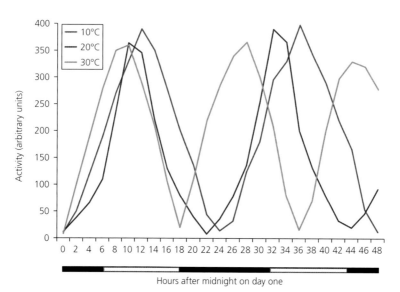

Figure 3.10.4   Activity rhythm of psychofish at temperatures of 10°C, 20°C and 30°C over 48 hours under otherwise standard laboratory conditions in the dark. Time 0 is midnight on day one, the start of the study. Note that before the experiment the fish were living in a strict 12-h light:12-h dark environment. The bar under the figure indicates when the lights would have been on (white) or off (black) if the 12:12 light:dark regime had been continued.

Question 2: What is the relationship, if any, between temperature and activity rhythm in psychofish? (5%)

Your initial data suggests that at 10°C the psychofish exhibit a 24-hour activity rhythm. However, further investigations reveal that this is not a universal observation in all 100 tanks. Interestingly, as the data in Fig. 3.10.5 show, populations of psychofish in 4 of the 100 tanks, specifically tanks 13, 17, 68 and 70, do not seem to show a 24-hour rhythm at 10°C.

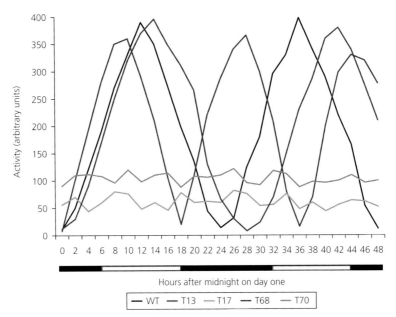

Figure 3.10.5 Activity rhythms of psychofish from tanks 13 (T13), 17 (T17), 68 (T68), 70 (T70) and a control wild-type (WT) population. These experiments were carried out at 10°C over 48 hours under otherwise standard laboratory conditions in the dark. Time 0 is midnight on day one, the start of the study. Note that before the experiment the fish were living in a strict 12-h light:12-h dark environment. The bar under the figure indicates when the lights would have been on (white) or off (black) if the 12:12 light:dark regime had been continued.

Question 3: How would you prove that the abnormal activity rhythm phenotypes found in psychofish from tanks 13, 17, 68 and 70 are genetic in origin? (5%)

Having established that the abnormal activity rhythm phenotypes are genetic in origin, you try to identify the gene(s) that, when mutant, will cause the activity rhythm phenotype. Fortunately, the Psychofish Genome Project has recently been completed and you decide to sequence the entire genomes of all 100 strains of psychofish to help you identify the gene(s) associated with abnormal rhythm.

Question 4: How would you use existing data from the Psychofish Genome Project, plus your new whole genome sequence data from the 100 tanks, to identify the gene(s), associated with the mutant phenotypes shown in the four strains? What problems could you encounter and how would you try to resolve them? (5%)

Your preliminary investigations suggest that the abnormal circadian rhythm phenotypes of the four mutant psychofish strains (13, 17, 68 and 70) are associated with mutations in a single gene, which you call *greenwich*. The precise mutations, as identified by DNA sequence analysis, are shown in Table 3.10.1.

Table 3.10.1   Nature of the four mutations found within the *greenwich* gene for each of the four mutant strains. The mutations in strains 13, 17 and 68 are found within the putative coding region. The mutation in strain 70 is within an intron.

Strain	Mutation
13	missense
17	nonsense
68	missense
70	insertion

Question 5: Speculate on how the different types of mutation listed in Table 3.10.1 might cause the specific rhythm phenotypes shown in Fig. 3.10.5. (5%)

You speculate that as the psychofish exhibit rhythmic behaviour associated with the *greenwich* gene, then the mRNA of the *greenwich* gene may also show a 24-hour oscillation. If this is true, then sequences upstream of the *greenwich* gene presumably drive expression of the *greenwich* gene in a 24-hour rhythmic pattern. You decide to investigate whether such a control sequence exists.

To investigate the control of *greenwich* expression, you fuse the wild-type *greenwich* gene control region to the coding sequence of green fluorescence protein (GFP). This allows you to create transgenic psychofish strains that are wild-type and in addition carry one of these transgenes. As psychofish are predominantly transparent, GFP activity can be measured in living transgenic psychofish. Including appropriate controls, you make three different *greenwich* transgenes, which are described in Fig. 3.10.6.

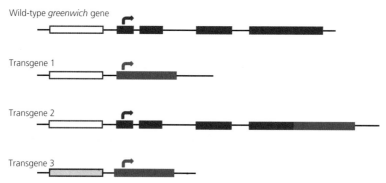

Figure 3.10.6   Cartoon images of the *greenwich* reporter genes used in this study. The wild-type *greenwich* gene comprises a control region (open brown box) and four exons (four closed brown boxes with the translational start site indicated by an arrow). Transgene 1 comprises the wild-type *greenwich* control region (open brown box) fused to the coding region of GFP (closed green box with translational start site indicted by an arrow). Transgene 2 comprises the wild-type *greenwich* control region (open brown box) and the first three and a half exons of the *greenwich* coding sequence (closed brown box with the translational start site indicated by an arrow) fused in frame to the coding sequence of GFP (closed green box). Transgene 3 comprises a constitutively expressed control region (open blue box) fused to the coding region of GFP (closed green box with translational start site indicated by an arrow).

You use standard, well-established molecular techniques to put the three transgenes into wild-type psychofish and generate several stable lines, each containing one of the transgenes. GFP expression patterns in representative individual psychofish carrying each class of transgene are shown in Fig. 3.10.7.

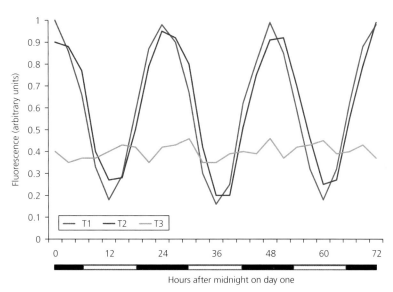

Figure 3.10.7   GFP expression in wild-type psychofish carrying one of the three transgenes, namely transgene 1 (T1), transgene 2 (T2) or transgene 3 (T3) described in Fig. 3.10.6. These experiments were carried out at 10°C over 72 hours under otherwise standard laboratory conditions in the dark. Time 0 is midnight on day one, the start of the study. Note that before the experiment the fish were living in a strict 12-h light:12-h dark environment. The bar under the figure indicates when the lights would have been on (white) or off (black) if the 12:12 light:dark regime had been continued.

Question 6: What should each of the three transgenes (T1, T2 and T3) reveal about the control of *greenwich* gene expression? In addition, when introduced to wild-type psychofish, why do each of the three transgenes show the circadian GFP expression pattern revealed in Fig. 3.10.7? (10%)

To further your understanding of *greenwich* gene control, you sensibly decide to investigate *greenwich* mRNA abundance in wild-type and homozygous *greenwich* null mutant psychofish. The *greenwich* null mutants you work with are from strain 17 described earlier and have a nonsense mutation in codon 6 of the *greenwich* gene. These data are shown in Fig. 3.10.8.

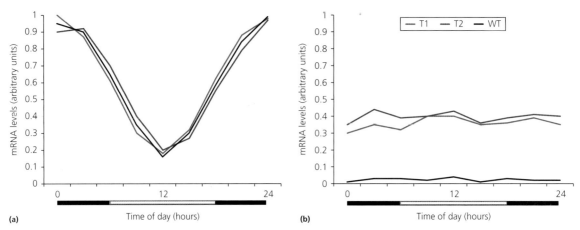

(a) Time of day (hours) (b) Time of day (hours)

Figure 3.10.8 mRNA levels of the endogenous *greenwich* gene (WT) and transgenes 1 (T1) and transgene 2 (T2) in (a) wild-type and (b) *greenwich* null mutant hosts. These experiments were carried out at 10°C over 24 hours under standard laboratory conditions in the dark. Time 0 is midnight on day one, the start of the study. Note that before the experiment the fish were living in a strict 12-h light:12-h dark environment. The bar under the figure indicates when the lights would have been on (white) or off (black) if the 12:12 light:dark regime had been continued.

Question 7: What is the significance of using a homozygous *greenwich* null mutant psychofish strain as a host for the transgenes? (5%)

Question 8: What do the data in Fig. 3.10.8 tell you about the control of *greenwich* gene expression in psychofish? (5%)

A few days later you are contacted by one of Professor Second's most talented former students, the newly promoted Professor Annie Versarrie. She has identified a second psychofish gene that she has shown to play a key role in 24-hour rhythmic behaviour. Professor Versarrie first found mutant psychofish quite by accident when she noticed that in one of the tanks the fish were very quiet whenever she arrived at work each morning. She called the mutant gene *lazy*, and later discovered that *lazy* mutants have a 30-hour rhythm.

In her studies, Professor Versarrie found another four psychofish strains showing abnormal or absent rhythms that seem to be caused by mutations in the *lazy* gene. Three of these mutant psychofish strains have no detectable rhythm and are called *lazy*^zero1, *lazy*^zero2 and *lazy*^zero3, whilst the forth strain has a short 18-hour rhythm and is called *lazy*^short.

You are puzzled by how different mutations in the same gene can generate such different phenotypes, so you use qPCR to measure *lazy* mRNA levels in wild-type and *lazy* mutant psychofish every 3 hours over a 2-day (48h) period. The data are shown in Table 3.10.2.

Table 3.10.2   Relative *lazy* mRNA levels (arbitrary units) in six strains of psychofish over a 48-h period. The six strains are wild-type and five independently derived mutant strains identified by Professor Versarrie and described earlier that carry mutations in the *lazy* gene. The original *lazy* mutant strain has a 30-h rhythm. Strains *lazy*^{zero1}, *lazy*^{zero2} and *lazy*^{zero3} have no detectable circadian rhythm. The *lazy*^{short} strain has an 18-h rhythm. Strains were initially grown in a 12-h light:12-h dark environment. At time 00.00 they were transferred to constant darkness. Under the original light/dark regime, lights would be switched on at 6 and off at 18, and so on.

Time (h)	Strains					
	wild-type	lazy^{zero1}	lazy^{zero2}	lazy^{zero3}	lazy	lazy^{short}
00.00	20	0	20	7	20	20
03.00	18	0	18	7	19	16
06.00	10	0	10	7	15	4
09.00	2	0	2	7	5	0
12.00	0	0	0	7	1	4
15.00	2	0	2	7	0	16
18.00	10	0	10	7	1	20
21.00	18	0	18	7	5	16
24.00	20	0	20	7	15	4
27.00	18	0	18	7	19	0
30.00	10	0	10	7	20	4
33.00	2	0	2	7	19	16
36.00	0	0	0	7	15	20
39.00	2	0	2	7	5	16
42.00	10	0	10	7	1	4
45.00	18	0	18	7	0	0
48.00	20	0	20	7	1	4

Question 9: Taking each of the five strains in turn, speculate on the type of mutation that could cause the observed *lazy* mRNA and circadian phenotypes. Discuss whether each of the mutations are likely to be dominant, co-dominant or recessive when heterozygous with the wild-type allele. (10%)

Question 10: The five different *lazy* mutations can generate 10 different transheterozygous combinations of two different *lazy* mutant alleles. Speculate on what their phenotypes might be. (5%)

After further discussions with Professor Second, you decide to look at levels of Greenwich and Lazy protein within psychofish over 24 hours. You calculate Greenwich and Lazy protein levels by isolating protein at various times during the 24-hour observation period, performing SDS polyacrylamide gel electrophoresis and then Western blotting using anti-Greenwich and anti-Lazy antisera. You also decide to investigate the effect of a light pulse on Greenwich and Lazy protein during the normal cycle. These data are shown in Fig. 3.10.9.

Figure 3.10.9   Western blots of Greenwich and Lazy protein in psychofish during a 24-hour day. Time 0 refers to midnight. Protein levels were calculated every 3 hours. Experiments were carried out at a constant 10°C in the dark. The red arrows show the time of the 30 minute light pulse. Note that before the experiment the fish were living in a strict 12-h light:12-h dark environment. The bar under the figure indicates when the lights would have been on (white) or off (black) if the 12:12 light:dark regime had been continued.

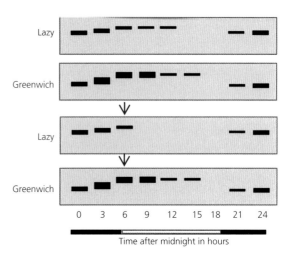

Time after midnight in hours

Question 11: What do the Western blot data tell you about the Greenwich and Lazy proteins during the normal 24-hour cycle? (5%)

Question 12: What do these data tell you about the effect of light pulses on the Greenwich and Lazy proteins? (5%)

You are interested in investigating precisely where in the cell the Greenwich and Lazy proteins are found. To address this problem, you generate a double mutant cell line that has no endogenous expression of the *greenwich* and *lazy* genes. Then, using standard and well-established techniques, you introduce a construct containing the *greenwich* and *lazy* coding regions under the control of a constitutively active promoter. You then try to determine where the Greenwich and Lazy proteins are found in the cell.

Question 13: In the cell culture experiments designed to investigate protein localisation, why were the *greenwich* and *lazy* genes expressed from a constitutively active promoter, rather than their own promoters? (5%)

You soon determine that when Greenwich protein alone was expressed, it was found to be in the cytoplasm. Similarly, when Lazy protein alone was expressed, it was also cytoplasmic. However, when Greenwich and Lazy were expressed together in the same cells, both proteins were found in the nucleus.

Preliminary amino acid sequence analysis reveals that both Greenwich and Lazy proteins have a nuclear localisation sequence. In particular, sequences between amino acids 88 and 575 of the Greenwich protein seem to inhibit nuclear localisation of the Greenwich protein, but the mechanism by which this happens is unclear.

Professor Second is keen for you to investigate the mechanism behind the nuclear localisation of the Greenwich protein and she provides you with a series of modified *greenwich* genes that will generate a set of deletion derivatives of the Greenwich protein. These derived Greenwich proteins, which all have specific tracks of amino acids missing, are shown in Fig. 3.10.10. These Greenwich deletion proteins were expressed in the cell culture described previously, to narrow down the amino acid(s) responsible for inhibition of nuclear localisation.

Question 14: Briefly explain how deletion mutants, such as those shown in Fig. 3.10.10, could be made. (5%)

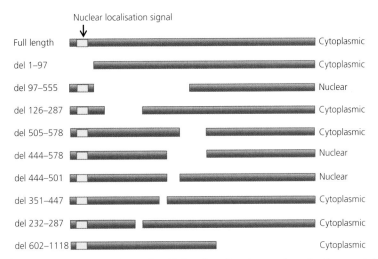

Nuclear localisation signal

Full length	Cytoplasmic
del 1–97	Cytoplasmic
del 97–555	Nuclear
del 126–287	Cytoplasmic
del 505–578	Cytoplasmic
del 444–578	Nuclear
del 444–501	Nuclear
del 351–447	Cytoplasmic
del 232–287	Cytoplasmic
del 602–1118	Cytoplasmic

Figure 3.10.10 Schematic representation of cellular localisations of each Greenwich deletion protein expressed, along with wild-type Lazy protein, in the cell culture described earlier. The amino acids deleted in each case are shown on the left. The figure shows the Greenwich amino acids retained in the protein. The solid box shows the position of the nuclear localisation signal, which is retained in 9 of the 10 Greenwich proteins. The exception is the Greenwich (del 1-95) protein in which the nuclear localisation sequence is deleted, so the protein is always located in the cytoplasm.

Question 15: Which sequence(s) are responsible for the cytoplasmic localisation of Greenwich protein? (5%)

Question 16: How could you characterise more precisely which amino acids are important for the cytoplasmic localisation of Greenwich protein? How could you approach this experimentally? (5%)

Professor Second is delighted with your progress and she is sure that you now have enough data to finally solve the problem of how psychofish know what time of day it is.

Question 17: Propose a model explaining how the *greenwich* and *lazy* genes regulate 24-hour rhythm in psychofish. How would you test your model? (10%)

Answers

Chapter 1.1: Grandma's Secret

The key to understanding the questions in this chapter is an understanding of how ABO blood groups and short tandem repeats are inherited. Details can be found in most genetics textbooks (such as *Principles of Genetics*, D. P. Snustad & M.J. Simons, John Wiley & Sons Ltd.) and reliable websites.

Question 1: What is the genetic basis of the ABO blood group system? (10%)

Answer 1: ABO blood group phenotypes are determined by three common alleles (I^A, I^B and I^O) that belong to a single gene called the *I* gene. The relationship between phenotype and genotype is shown in the following table.

Phenotype	Genotype
blood group A	$I^A I^A$ or $I^A I^O$
blood group B	$I^B I^B$ or $I^B I^O$
blood group O	$I^O I^O$
blood group AB	$I^A I^B$

In this case, the I^A and I^B alleles are co-dominant and, in addition, are both dominant to the I^O allele. The biochemical reason for this is quite simple as the I^A allele encodes the A version of the blood group cell-surface protein and the I^B allele encodes the slightly different B version of the blood group cell-surface protein, whilst the I^O allele doesn't encode anything, or at least doesn't encode a functional protein that can sit on the cell surface. The three ABO alleles are inherited in a standard Mendelian fashion, in that you receive one allele from your mother and one allele from your father.

Question 2: Briefly discuss the limitations of testing relatedness using ABO blood groups. (10%)

Answer 2: Working on the assumption that you only know the ABO phenotypes (as would have been the case in the 1960s), then a parent of blood group AB cannot have a child of blood group O, and *vice versa*. All other combinations are possible. In the 1960s, paternity testing of this type

Genetics? No Problem!, First Edition. Kevin O'Dell.
© 2017 John Wiley & Sons Ltd. Published 2017 by John Wiley & Sons Ltd.

could only exclude individuals from being the genetic father but could not be used to absolutely confirm parentage. ABO data was usually used in conjunction with known inherited variation that could be measured biochemically and that was known to be associated with alleles at a single gene. This strategy was used to narrow down the number of suspects in the first forensics case using the new technique of DNA fingerprinting, which was developed by Sir Alec Jeffreys in the 1980s.

Question 3: What could the genotypes of your Grandma, Paul Cool and Aunt Brenda be, if your Grandma and Paul Cool are indeed Aunt Brenda's parents? (10%)

Answer 3: As the child is blood group A and the alleged father is blood group B, then the only combination that works is if the child receives an I^O allele from the alleged father, as shown in the following table.

Subject	Blood group	Genotype
mother (your grandma)	AB	$I^A I^B$
child (Aunt Brenda)	A	$I^A I^O$
alleged father (Paul Cool)	B	$I^B I^O$

Question 4: What could the genotypes of your Grandma, Paul Cool and Aunt Brenda be if your Grandma is Aunt Brenda's mother but Paul Cool is not Aunt Brenda's father? (10%)

Answer 4: If either the child or the alleged father is homozygous (the child being $I^A I^A$ or the alleged father being $I^B I^B$) then they cannot be first-degree relatives.

Subject	Blood group	Genotype
mother (your Grandma)	AB	$I^A I^B$
child (Aunt Brenda)	A	$I^A I^A$ or $I^A I^O$
alleged father (Paul Cool)	B	$I^B I^B$ or $I^B I^O$

Of course in reality both could be heterozygous (the child being $I^A I^O$ and the alleged father being $I^B I^O$) and still not be related as so many people have those genotypes. In other words, ABO phenotype data cannot prove that someone is the parent of a specific child. It can only be used for exclusion.

Question 5: Why is Dr Robert's interpretation correct? (5%)

Answer 5: As you know both the genotype of the mother ($I^A I^B$) and that the child has blood group A, the child must inherit an I^A allele from her mother and either an I^O or I^A allele from her real genetic father. As the alleged father may carry an I^O allele, you cannot exclude the possibility that he is the genetic father of the child. This illustrates the limitations of using ABO blood groups in paternity disputes, as only AB parents of O children (and *vice versa*) are impossible.

Question 6: What are short tandem repeats (STRs)? (10%)

Answer 6: A STR is a short repetitive sequence of base pairs. Each STR is usually 2, 3 or 4 base pairs long. For example, if the core of the repetitive sequence is TAG, then an STR with 5 repeats would be TAGTAGTAGTAGTAG. Within a species, every individual has the same STR at the same position on the chromosome. However, individuals vary in the number of repeats they have. So using the TAG repeat as an example, there may be several different TAG repeat numbers in the human population such as 5 repeats (TAGTAGTAGTAGTAG), 6 repeats (TAGTAGTAGTAGTAGTAG), 7 repeats (TAGTAGTAGTAGTAGTAGTAG), and so on.

Question 7: Briefly describe an experiment that would enable you to determine the length of an STR. (10%)

Answer 7: For humans and many other organisms, the chromosomal position and sequence of STRs are well documented. Therefore, for the STR used in this example you will know the sequence of the DNA flanking the repeat. You can therefore design primers that recognise the sequence flanking the repeat and use polymerase chain reaction (PCR) to amplify the sequence between the two flanking sites. As a result, you will amplify two fragments of DNA (one from each chromosome) whose sizes reflect the number of STR repeats. You can then separate these DNA fragments according to size on an electrophoretic gel.

Question 8: Why are STRs used to determine relatedness, especially paternity? (5%)

Answer 8: STRs are inherited in a Mendelian fashion. In other words each STR is present in two copies, one on the chromosome you inherit from your mother and the other from the chromosome that you inherit from your father. Again you pass on one of each of your STRs to each of your children. In theory any piece of DNA can be used to study relatedness, but STRs are particularly useful as they are very variable.

Question 9: Who is Aunt Brenda's genetic father? (10%)

Answer 9: At this STR your Grandma has the genotype 8/17, whilst Aunt Brenda's genotype is 8/11. Therefore Aunt Brenda receives the 8 STR allele from your Grandma and the 11 STR allele from her genetic father. The limited data you have exclude Paul Cool (14/16) as the father and suggest your Grandad (11/19) is in fact Aunt Brenda's father.

Question 10: Draw a family tree to show the apparent relationship between everyone in Table 1.1.2. (10%)

Answer 10: Family tree as shown.

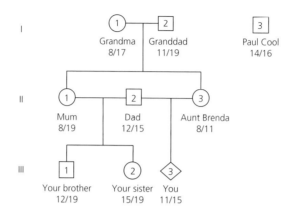

Question 11: What are you going to tell your mother? (5%)

Answer 11: It would seem that your genetic parents are in fact your father and your Aunt Brenda (not your 'mother'). Presumably your 'mother' knows this, but she may not expect you to know this.

Chapter 1.2: Tiger! Tiger!

Question 1: Why is inbreeding often a problem for wild animals? (10%)

Answer 1: Most deleterious mutations are recessive, so as long as an individual is heterozygous, it's normally not a problem for its fitness, health, viability and fecundity. In inbred populations, however, an individual is very likely to be homozygous for one or more of these deleterious recessive mutations and, as a result, its fitness, health, viability or fecundity is likely to be affected.

Question 2: What kind of plan could conservation project leaders, such as Professor Bagh, follow to avoid issues with inbreeding? (10%)

Answer 2: Outbreeding. If DNA testing (or any other route) can identify relatedness, then it should be possible to only allow unrelated individuals to mate and produce offspring.

Note that in tigers and many other species, there is a key conservation question in deciding whether different populations of a species are so different that the populations themselves are valuable in their own right. Therefore crossing between different populations might be considered undesirable for retaining the identity of a specific population, but may well be desirable for maintaining the genetic diversity and viability of the species. There is no simple answer to this.

Question 3: Looking at the image of the mutant white tiger in Fig. 1.2.1, give two reasons why the white Bengal tiger is not a true albino. (10%)

Answer 3: The white tiger has black stripes, so can clearly make melanin. In addition, it does not have pink/red eyes. True albinos have no significant melanin pigmentation and have pink/red eyes.

Question 4: What is the evidence that the white colour is caused by a recessive mutation? (10%)

Answer 4: The data in Fig. 1.2.3 show that white tigers can have two orange parents. If the white colour was from a dominant allele this would not be possible.

Question 5: Redraw the pedigree showing the genotypes of each individual. (10%)

Answer 5: Using the notation + for the dominant orange allele and − for the recessive white allele, you can redraw the pedigree with genotypes as shown in this family tree.

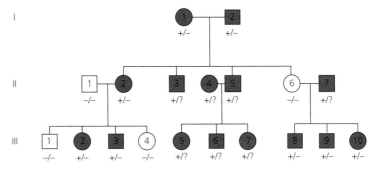

Question 6: There may be some individuals whose genotypes cannot be absolutely determined. Who are they and why can't their genotypes be determined? (10%)

Answer 6: Orange individuals II:3 and II:5 have heterozygous (+/− × +/−) parents, and could themselves therefore be homozygous (+/+) or heterozygous (+/−). As individuals III:5, III:6 and III:7 may also be derived from two heterozygous parents, their genotypes are similarly unknown. The parents of II:4 and II:7 are unknown and they may be homozygous or heterozygous. Note that the numbers here are too small to allow any inference from the ratios of phenotypes in the offspring, and similarly do not allow for any statistical analysis.

Question 7: How could you determine the precise genotypes of the problem animals identified in your answer to question 6? (10%)

Answer 7: The genotypes of individuals that are phenotypically orange but are of unknown (either +/+ or +/−) genotype could be distinguished by mating them to a homozygous white (−/−) tiger. If the orange parent is homozygous (+/+) then all their offspring would be orange (+/−), whereas if the parent is heterozygous (+/−) half their offspring would be orange (+/−) and half would be white (−/−). If you knew the molecular basis of the mutation (the specific DNA change) you could design a DNA test to see whether the individual animals are homozygous (+/+) or heterozygous (+/−) orange.

Question 8: What is the genetic basis of STRs, and how can they be used to determine parentage? (10%)

Answer 8: A short tandem repeat (STR) is a short repetitive sequence of base pairs. They are usually 2, 3 or 4 base pairs long. For example if the core of the repetitive sequence is TAG then an STR with 5 repeats would be TAGTAGTAGTAGTAG. Within a species every individual has the same STR at the same position on the chromosome. However they vary in the number of repeats. STRs are inherited in a Mendelian fashion, meaning that every individual has two 'alleles' of each STR, and inherit one from their mother and one from their father.

Question 9: Who are the parents of the three orange cubs: Clementine, Satsuma and Tangerine? (10%)

Answer 9: Taking each cub in turn:
Clementine receives the 27 allele from Sunetra and the 31 allele from Sourav.
Satsuma receives the 22 allele from Sunetra and the 31 allele from Sourav.
Tangerine receives the 27 allele from Sunetra and the 34 allele from Sourav.

Question 10: What is the most plausible explanation for Clementine, Satsuma and Tangerine being orange? (10%)

Answer 10: Sunetra and Sourav are white tigers, yet their offspring are all orange. The most plausible explanation is that Sunetra and Sourav have mutations in different genes, meaning that their cubs are heterozygous at both genes.

Let us suppose that Sunetra is white because she is homozygous for a recessive mutation in the gene that you've already identified. You can call the gene W and say that the dominant wild-type orange allele is w^+ and the recessive white mutant allele is w^-, meaning that Sunetra has the genotype w^-w^-. Similarly, Sourav is homozygous for a recessive mutation in a different gene that you could call T, where the domain orange allele is t^+ and the recessive white mutant allele is t^-, meaning Sourav has the genotype t^-t^-. The full genotypes of the two tiger parents would therefore be, Sunetra ($t^+t^+\ w^-w^-$) and Sourav ($t^-t^-\ w^+w^+$). Their three offspring would have the genotypes ($t^+t^-\ w^+w^-$) and, as they have a wild-type allele at each gene, they will be orange.

Geneticists call this phenomenon, complementation.

Further reading

See http://www.bbc.co.uk/news/science-environment-24117279 (accessed July 2015).

Chapter 1.3: Anticipation

Question 1: How is Huntington's disease inherited? (10%)

Answer 1: It is an autosomal dominant condition.

Question 2: What is the molecular basis (DNA change) of the mutation that causes Huntington's disease? (10%)

Answer 2: It is caused by an expanded (CAG) trinucleotide repeat within the coding region of the Huntington gene. As CAG encodes glutamine, the resulting protein has a polyglutamine track, the length of which reflects the number of CAG codons. Generally the longer the repeat, the earlier the age of onset and the more severe the condition.

Question 3: Briefly describe the phenotype of Huntington's disease. (10%)

Answer 3: It is a progressive neurodegenerative condition. In particular, it affects muscle coordination and leads to a decline in mental capacity. The first symptoms often manifest as subtle problems with mood or cognition and can often be mistaken for other conditions. This is followed by progressive lack of coordination and as the disease advances, uncoordinated and jerky body movements become more apparent and are coupled with a decline in mental ability that ultimately declines into dementia.

Question 4: How would you obtain the DNA from each family member? (10%)

Answer 4: The normal route would be to use buccal swabs or blood samples, though any cell or tissue material could be the source of the DNA.

Question 5: Briefly describe, using an appropriate figure if necessary, the process by which the DNA test is undertaken, as depicted in the image in Fig. 1.3.1. (20%)

Answer 5: Use polymerase chain reaction (PCR). Primers should recognise the unique DNA sequences flanking the CAG repeat of the Huntingtin gene, described in the answer to question 2, to amplify the DNA. The larger the repeat, the bigger the amplified sequence of DNA. DNA on the gel separates according to size, so small (unaffected fragments) travel further. Of course it would also be possible simply to sequence the DNA, but that doesn't fit the data as presented in Fig. 1.3.2.

Question 6: Which members of the family have Huntington's disease? What is the evidence for this? (30%)

Answer 6: Freddie, Agnes and Edward have Huntington's disease. They each have a long (disease-causing) fragment of DNA containing an expanded CAG repeat, specifically a CAG repeat of 46 (Freddie), 52 (Agnes) and 54 (Edward). The CAG fragment gets longer with succeeding generations, a process called anticipation.

Question 7: How reliable is this diagnostic test for Huntington's disease? (10%)

Answer 7: The test is very reliable, especially as the repeat size determines whether or not someone has the condition. In addition, the CAG repeat length primarily – but not absolutely – determines the age of onset of the condition.

Chapter 1.4: Budgie Hell

Question 1: What is the most plausible reason for the white budgie female having a clutch of seven chicks all of which are blue? (10%)

Answer 1: As Mrs Singh has kept her budgies in aviaries comprising budgies of specific colours and is surprised when chicks of a different colour appear in a nest, you can be fairly confident that the budgies within each aviary are true-breeding, meaning that they are homozygous for any genes associated with colour. Therefore, for a white female to have blue chicks she must have mated with a blue male. The simplest explanation for the chicks being blue is that there is a gene that determines whether the budgie is blue or white and that the blue allele is dominant to the white allele.

Question 2: From a genetic perspective, why isn't it usually a good idea for siblings to have offspring? (10%)

Answer 2: Most mutations are recessive and many recessive mutations are deleterious. As most recessive mutations are rare, it follows that inbred populations are more likely to be homozygous for recessive mutations than outbred populations, and therefore inbred individuals are likely to be less fit than outbred individuals. As siblings share about 50% of their genes, there is a high chance that they would produce offspring that are homozygous for a deleterious recessive mutation.

Question 3: What is the most plausible genetic explanation for the colours of the chicks in Aunt Polly's aviary? (10%)

Answer 3: These data support the answer to question 1, as the F2 show a 3:1 ratio. The most plausible explanation is that there is a single gene associated with the blue/white colour and the blue allele is dominant to the white allele. Using the notation b^+ for the dominant blue allele and b^- for the recessive white allele, you can construct a Punnett square:

		Paternal allele	
		b^+	b^-
Maternal allele	b^+	b^+b^+ blue	b^+b^- blue
	b^-	b^+b^- blue	b^-b^- white

Question 4: Using a chi-squared test, determine whether your explanation of the inheritance of blue and white colours in budgies is likely to be correct. (10%)

Answer 4: Using the table provided:

Colour phenotype	Observed number	Expected number	$(O - E)$	$(O - E)^2$	$\dfrac{(O-E)^2}{E}$
blue	25	24	1	1	0.04
white	7	8	-1	1	0.13
total	32	32			0.17

As the chi-squared value of 0.17 is less than 3.84, you accept your hypothesis. There is a single gene segregating here, where the blue allele is dominant to the white allele.

Question 5: What is the most plausible genetic explanation for the colours of the chicks in Aunt Robyn's aviary? (5%)

Answer 5: Once again it would seem that you've identified a single gene that is associated with a yellow colour. The simplest explanation is that at this gene the yellow allele is dominant to white. Upon mating the heterozygous (yellow/white) F1 yellow birds together, you would expect to generate an F2 in a 3:1 ratio. The numbers of F1 individuals you see are 38:10. Using the notation y^+ for the dominant yellow allele and y^- for the recessive white allele, you can construct a Punnett square:

		Paternal allele	
		y^+	y^-
Maternal allele	y^+	y^+y^+ yellow	y^+y^- yellow
	y^-	y^+y^- yellow	y^-y^- white

Question 6: Using a chi-squared test, determine whether your explanation of the inheritance of yellow and white colours in budgies is likely to be correct. (5%)

Answer 6: Constructing your own table:

Colour phenotype	Observed number	Expected number	(O − E)	(O − E)^2	$\dfrac{(O-E)^2}{E}$
yellow	38	36	2	4	0.11
white	10	12	-2	4	0.33
Total	48	4			0.44

As before, there is one degree of freedom. The chi-squared value of 0.44 is less than 3.84, so you accept your hypothesis. There is a single gene segregating here, where the yellow allele is dominant to the white allele.

Question 7: What is the most plausible genetic explanation for the colours of the chicks in Aunt Phoebe's aviary? (10%)

Answer 7: This looks like a 9:3:3:1 ratio, immediately suggesting that two genes are involved. Working with the idea that the simplest explanation that works is usually correct, it's likely that it's the same two genes you've already met (blue vs white and yellow vs white), and when the budgie has a dominant blue allele at the blue/white gene and a dominant yellow allele at the yellow/white gene, they appear green.

Question 8: Using a chi-squared test, determine whether your explanation of the inheritance of colour in Aunt Phoebe's budgies is likely to be correct. Again there will be 1 degree of freedom and the 5% cut-off for significance is 4.85. (10%)

Answer 8: Constructing your own table:

Colour phenotype	Observed number	Expected number	$(O - E)$	$(O - E)^2$	$\dfrac{(O - E)^2}{E}$
green	224	225	−1	1	0.004
blue	83	75	8	64	0.853
yellow	70	75	−5	25	0.333
white	23	25	−2	4	0.160
Total	400	400			1.350

As there is one degree of freedom, and the chi-squared value of 1.35 is less than 4.85, you accept your hypothesis. There are two genes segregating here, where the yellow allele is dominant to the white allele, and at a second gene the blue allele is dominant to the white allele. When wild-type at both genes, the blue and yellow birds are green.

Question 9: Speculate on the type of proteins the gene(s) might encode. (15%)

Answer 9: Genes encode proteins, so in this case what you really want to know is what proteins could generate the colours seen in the budgies. As you are asked to 'speculate', any plausible answer is correct.

There are two genes, one associated with blue/white (where blue is dominant) and the other with yellow/white (where yellow is dominant). So the most plausible start point is to think that the wild-type allele at the blue/white gene 'makes' blue and the wild-type allele at the yellow/white gene 'makes' yellow.

Colour is often associated with pigments, but pigments are not proteins, so genes cannot make (encode) pigments directly. But genes can encode proteins (enzymes) that can catalyse the conversion of one pigment to another. So a plausible answer is that the blue/white gene encodes an enzyme that catalyses the conversion of a white pigment to a blue pigment, whilst the yellow/white gene catalyses the conversion of a white pigment to a yellow pigment.

Note that there are other equally plausible and equally correct answers to this question.

Question 10: Propose a genetic explanation for the colour pattern of *Harlequin*. (15%)

Answer 10: Harlequin is a male and on first inspection it looks as if the left-hand-side of his body makes both yellow and blue pigment (as it is green) and the right-hand-side can only make blue pigment (as it is blue). So the issue you need to address is why half of Harlequin makes yellow pigment and half doesn't.

The most plausible explanation would be that a new mutation arose during his early development. As mutations are rare, a good explanation would require the fewest mutations possible. In addition, most mutations are loss of function, so the most plausible explanation would be that the embryo starts with the capacity to make yellow pigment everywhere but one cell in early development loses that ability. To go from 'making yellow' to 'not making yellow' in a single mutation, Harlequin must start life as a heterozygous (y^+y^-) fertilized egg, and by chance acquire a mutation in the y^+ allele in one cell, so that cell is now y^-y^-, and all daughter cells derived from that mutated cell are also y^-y^-. Given the pattern of colours in Harlequin, it is tempting to think this occurred very early in development.

Again, note that there are other equally plausible and equally correct answers to this question.

Chapter 1.5: Friends Reunited

Question 1: Define the term 'polymorphism'. (10%)

Answer 1: In their book *A Dictionary of Genetics*, R.C. King, W.D. Stansfield and P.K. Mulligan (Oxford University Press) define polymorphism as 'The existence of two or more genetically different classes in the same interbreeding populations. The polymorphism may be transient, or the proportions of the different classes may remain the same for many generations. In the latter case the phenomenon is referred to as a balanced polymorphism'.

Similarly D.P. Snustad and M.J. Simmons in their text *Principles of Genetics* (John Wiley & Sons, Ltd) use the definition, 'The existence of two or more variants in a population of individuals, with at least two of the variants having frequencies of greater than 1 percent'.

Question 2: For studies of this type, what are the advantages of looking at non-coding, rather than coding, sequences? (15%)

Answer 2: Coding sequences, or in fact any sequence that has function, is under the influence of selection. Therefore, whilst new mutations arise in these sequences at a standard rate, the probability that they are retained depends primarily on whether they confer an advantage, a disadvantage or are selectively neutral. Thus, although mutations in coding DNA arise at a constant rate, they are not retained at constant rate. Mutations in non-coding DNA (or perhaps more correctly in DNA that has no function) will arise at a constant rate and be retained at a constant rate, so the number of differences between two non-coding (or non-selected) sequences reflects the time since they diverged.

Question 3: Which students have identical DNA sequences at all 10 polymorphic sites? What does this tell you? (10%)

Answer 3: The 11 students have six different mitochondrial sequences namely:
Lucia: ACGCTCTCAA
Indira: ACTCTCATAA
Li & Zhang: ACTCTCTCAA
Ahmed & Farah: TATGTGACAA
Josh, Pablo & Sofia: TCTCTGACGA
Celia & Nicky: TCTCTGCCGA
As Li & Zhang, Ahmed & Farah and Josh, Pablo & Sofia, and Celia & Nicky have identical DNA sequences at the 10 bases, it would suggest that they are relatively closely related down their maternal lineages.

Question 4: Define the term 'haplotype' and calculate how many different mitochondrial DNA haplotypes can be seen can be found in your class. (15%)

Answer 4: The term 'haplotype' is a contraction of 'haploid genotype'. It is used to describe a combination of closely linked alleles or sequence polymorphisms that are inherited together. As given in the answer to question 3, there are six different mitochondrial haplotypes in the class.

Question 5: Which haplotypes differ by a single base-pair change? Can you determine within which line the DNA mutation has occurred? (15%)

Answer 5: Josh, Pablo & Sofia (TCTCTGACGA) differ from Celia & Nicky (TCTCTGCCGA) at the 7th base (A vs C). The historical sequence at the 7th base is A. Therefore the likeliest explanation is that the common maternal ancestor of the five students had the sequence (TCTCTGACGA) and the A in position 7 mutated to a C in the maternal lineage to Celia & Nicky.

Similarly, Lucia (ACGCTCTCAA) differs from Li & Zhang (ACTCTCTCAA) at the 3rd base (G vs T). The historical sequences at the 3rd base is T. Therefore, the likeliest explanation is that the common maternal ancestor of the three students had the sequence (ACTCTCTCAA) and the T in position 3 mutated to a G in the maternal lineage to Lucia.

Question 6: Construct an evolutionary tree that includes all 11 of the students in your class. Clearly show where each DNA base-pair change is likely to have occurred. (25%)

Answer 6: The evolutionary tree should look like this:

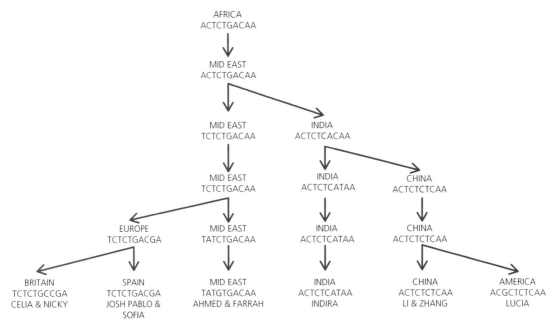

You could place this on a human evolutionary map as follows:

Question 7: Three of the students, Josh, Lucia and Sofia, are from the Americas. What do their mitochondrial DNA sequences reveal about their maternal ethnic origins? (10%)

Answer 7: Lucia's mitochondrial DNA is most closely related to that of the Chinese students Li & Zhang, so she is probably maternally descended from Native Americans. Josh & Sofia have the same mitochondrial DNA sequence as Pablo from Spain, so they are probably maternally descended from Hispanics.

Chapter 1.6: The Footballer, his Wife, their Kids and her Lover

Question 1: What is the genetic basis of STRs? (10%)

Answer 1: A short tandem repeat (STR) is a short repetitive sequence of base pairs. The repeat units are usually 2, 3 or 4 base pairs long. For example if the core of the repetitive sequence is TAG then an STR with 5 repeats would be TAGTAGTAGTAGTAG. Within a species every individual has the same STR at the same position on the chromosome. However, individuals vary in the number of repeats at each STR.

Question 2: In paternity disputes such as this, why are STR analyses more reliable than those using single nucleotide polymorphisms (SNPs) or other types of DNA variation? (15%)

Answer 2: When investigating relatedness, the more data you have to look at, the more reliable your analysis will be. Therefore, within the population, the more variable the sequence you are looking at, the more reliable your analysis will be. SNPs can only have four possible variants (A C G T), though in reality in most situations in most populations, most SNPs occur as alternate alleles and very few SNPs ever have all four possibilities segregating within it. Indeed, even where SNPs exist, the second most frequent allele is usually very rare. STRs are much better when investigating relatedness as they are much more variable.

Question 3: Using a clean and uncontaminated DNA sample as your start point, briefly describe the experimental strategy that culminates in the image of two bands in each lane as shown in Fig. 1.6.2. (20%)

Answer 3: For humans and many other organisms, the chromosomal position and sequence of STRs are well documented. Therefore, for the STR used in this example, you will know the sequence of the DNA flanking the repeat. You can therefore design primers that recognise the sequence flanking the repeat and use polymerase chain reaction (PCR) to amplify the sequence between the two flanking sites. As a result you will amplify two fragments of DNA (one from each chromosome) whose sizes reflect the number of STR repeats. You can then separate these DNA fragments according to size on an electrophoretic gel. This is precisely what is shown in Fig. 1.6.2.

Question 4: According to the data shown in Fig. 1.6.2, who are the parents of baby Frank? (15%)

Answer 4: Tracy and Percy Nutmeg.

Question 5: Are there any other issues in the data that Ed might be interested in? (20%)

Answer 5: Whilst Tracy is the mother of all five boys, Percy is the father of just three, as he is neither the father of Jacob nor Noah. They are not the sons of Franco either. In fact, Jacob and Noah seem to have the same father, but his identity is unknown.

Question 6: Ed is concerned about the reliability of the data. What further experiments could you do to reassure him? (20%)

Answer 6: Ed is right to be concerned. Using just one STR is potentially unreliable, as STRs can mutate, though this is quite rare. Therefore, it would be better to repeat this analysis with multiple STRs. Investigations into paternity disputes or other forensic analyses usually use 13 or more STRs simultaneously.

Chapter 1.7: Give Peas a Chance

Gregor Mendel was a contemporary of Bunsen and Nestler, and the pea data included here comes from Mendel's own experiments. Nestler investigated inheritance in sheep, though the Tsurcana data included here comes from another source.

Question 1: Try to improve Mendel's mood by explaining why the F2 plants are inherited in the ratios in Table 1.7.1. (10%)

Answer 1: The number of round (5474) and wrinkled (1850) pea plants in the F2 occurs at a ratio of 2.96 to 1. This is remarkably close to the standard Mendelian ratio of 3:1. Therefore the best explanation is that seed shape is determined by a single gene, with two alternative alleles, round and wrinkled. As the F1 are all round, then the round allele (r^+) is dominant to recessive wrinkled allele (r^-).

You are told that the parents are true-breeding round (r^+r^+) and true-breeding wrinkled (r^-r^-), so crossing them would generate F1 (r^+r^-) plants that are round, as round is the dominant allele.

If you cross two F1 plants together ($r^+r^- \times r^+r^-$)) then the F2 plants will have four possible genotypes, three of which (r^+r^+, r^+r^- and r^-r^+) are round and one (r^-r^-) is wrinkled.

Question 2: Explain to Mendel why crossing F1 round green pea plants to each other, gives F2 generation pea plants in the ratio 9 round green, 3 wrinkled green, 3 round yellow and 1 wrinkled yellow. (15%)

Answer 2: You can use the same notation as before (r^+ for the dominant round allele and r^- for the recessive wrinkled allele). The pattern of inheritance of the green/yellow colour is exactly the same as that of the round/wrinkled shape but in this case the green allele (g^+) is dominant to the recessive yellow allele (g^-).

If you cross wrinkled green plants ($r^-r^- \ g^+g^+$) to round yellow ($r^+r^+ \ g^-g^-$) plants, then all the F1 will have the genotype $r^+r^- \ g^+g^-$ and show the dominant characteristics round and green.

If you cross two F1 plants together ($r^+r^- \ g^+g^- \times r^+r^- \ g^+g^-$) then the F2 plants could have any one of 16 genotypes that segregate in the 9:3:3:1 ratio as shown in this Punnett square:

		F1 gametes			
		$r^+ \ g^+$	$r^+ \ g^-$	$r^- \ g^+$	$r^- \ g^-$
F1 gametes	$r^+ \ g^+$	$r^+r^+ \ g^+g^+$ round green	$r^+r^+ \ g^+g^-$ round green	$r^+r^- \ g^+g^+$ round green	$r^+r^- \ g^+g^-$ round green
	$r^+ \ g^-$	$r^+r^+ \ g^+g^-$ round green	$r^+r^+ \ g^-g^-$ round yellow	$r^+r^- \ g^+g^-$ round green	$r^+r^- \ g^-g^-$ round yellow
	$r^- \ g^+$	$r^+r^- \ g^+g^+$ round green	$r^+r^- \ g^+g^-$ round green	$r^-r^- \ g^+g^+$ wrinkled green	$r^-r^- \ g^+g^-$ wrinkled green
	$r^- \ g^-$	$r^+r^- \ g^+g^-$ round green	$r^+r^- \ g^-g^-$ round yellow	$r^-r^- \ g^+g^-$ wrinkled green	$r^-r^- \ g^-g^-$ wrinkled yellow

Question 3: Would Mendel get the same results if he'd started his first cross between pure-breeding round green pea plants and pure-breeding wrinkled yellow pea plants, rather than the cross between pure-breeding round yellow and wrinkled green as described above? (10%)

Answer 3: It doesn't matter what the combination of phenotypes is as the F1 is exactly the same. For example:

If you cross wrinkled green plants ($r^-r^-\,g^+g^+$) to round yellow ($r^+r^+\,g^-g^-$) plants, then all the F1 will have the genotype $r^+r^-\,g^+g^-$ and show the dominant characteristics round and green.

If you cross round green plants ($r^+r^+\,g^+g^+$) to wrinkled yellow ($r^-r^-\,g^-g^-$) plants, then all the F1 will have the genotype $r^+r^-\,g^+g^-$ and show the dominant characteristics round and green.

As the F1 from the two crosses have identical genotypes ($r^+r^-\,g^+g^-$), it follows that the F2 will have identical genotypes too.

Question 4: What classes and ratios of offspring would Mendel expect to see in the F2 if he were to cross a pure-breeding tall round green strain to a pure-breeding short wrinkled yellow strain? (15%)

Answer 4: The phenotypes will segregate in a 27:9:9:9:3:3:3:1 ratio. The relationships between the phenotypes and genotypes are shown in the following table. For the gene associated with height it is clear that tall (t^+) is dominant to short (t^-). In the table, the notation '?' means that it could be a dominant (t^+ or r^+ or g^+) or recessive (t^- or r^- or g^-) allele at that gene.

Ratio	Phenotype	Genotype
27	tall round green	T? R? G?
9	tall round yellow	T? R? gg
9	tall wrinkled green	T? rr G?
9	short round green	tt R? G?
3	tall wrinkled yellow	T? rr gg
3	short round yellow	tt R? gg
3	short wrinkled green	tt rr G?
1	short wrinkled yellow	tt rr gg

Question 5: How is coat colour inherited in Tsurcana sheep? (25%)

Answer 5: Taking each cross in turn:

black × black generates all black – this simply tells us that black individuals are homozygous;

black × grey generates 50% black and 50% grey – if black is homozygous, then to get a 50/50 ratio the simplest explanation is that grey is heterozygous;

grey × grey generates 67% grey and 33% black – the first thing to note is that the offspring are apparently in a 2:1 ratio, which is not a normal genetic ratio.

Remembering that the evidence from the earlier crosses suggested that grey is heterozygous and black is homozygous, you can determine the grey genotype as c^+c^- and black genotype as c^-c^-. The first two crosses make sense with this notation. If you then cross two greys together ($c^+c^- \times c^+c^-$)

the offspring would be c^-c^- (black), c^+c^- (grey), c^-c^+ (grey) and c^+c^+ (unknown), which would seem to be missing from the above cross. The only explanation is that the c^+c^+ genotypes are inviable and are never born, which is why you only see a 2 grey (c^+c^- and c^-c^+) and 1 black (c^-c^-) ratio.

Therefore you can describe the grey allele c^+ as dominant for colour (with the black allele c^- recessive for colour), but the c^+ allele is also recessive for lethality. Therefore an allele can be dominant for one characteristic but recessive for another.

Question 6: Why might the coat colour gene you've identified in Tsurcana sheep also have a role in another characteristic? (25%)

Answer 6: As it is difficult to explain why a mutant colour phenotype causes lethality in unborn sheep, but as colour in an unborn sheep is unlikely to affect viability directly, the mutation must affect another characteristic.

Chapter 1.8: Noah's ARC

Question 1: Speculate on how cream, light brown and white body colour is inherited in guinea pigs. (20%)

Answer 1: Whenever you see a cross that results in an F1 with a 1:2:1 ratio, where the '2' class of the F1 has the same phenotype as the parents (as in the final cross in Table 1.8.1), you should think that both original parents are heterozygous for two different alleles. In this example, suppose that the colour gene C has two distinct alleles white (C^w) and light brown (C^b). Individual guinea pigs can be homozygous white (C^w/C^w), homozygous light brown (C^b/C^b) or heterozygous cream (C^w/C^b). A situation where the heterozygous phenotype (in this case cream) is midway between the two homozygous parental phenotypes (in this case white and light brown) is called co-dominance, intermediate dominance or incomplete dominance. The precise term used depends on exactly why the heterozygote has an intermediate phenotype.

Note that other notation is perfectly acceptable, but it must be clearly defined. However, it is always better to make it clear that two alleles belong to the same gene, so B (brown) and b (white) is a reasonable alternative but implies that light brown is dominant to white. Note that B (brown) and W (white) is not appropriate as it clearly suggests the alleles are from different genes.

Question 2: Using appropriate genetic notation, redraw each of the crosses shown in Table 1.8.1 showing the genotypes of the parents and their offspring. (20%)

Answer 2: Considering each cross in turn, and using the notation defined earlier:

$$\text{white}\left(C^w/C^w\right)\times\text{light brown}\left(C^b/C^b\right)=100\%\text{ cream}\left(C^w/C^b\right)$$

$$\text{white}\left(C^w/C^w\right)\times\text{white}\left(C^w/C^w\right)=100\%\text{ white}\left(C^w/C^w\right)$$

$$\text{light brown}\left(C^b/C^b\right)\times\text{light brown}\left(C^b/C^b\right)=100\%\text{ light brown}\left(C^b/C^b\right)$$

$$\text{cream}\left(C^w/C^b\right)\times\text{cream}\left(C^w/C^b\right)=$$

$$25\%\text{ white}\left(C^w/C^w\right)+25\%\text{ light brown}\left(C^b/C^b\right)+50\%\text{ cream}\left(C^w/C^b\right)$$

You might want to draw a Punnett square to prove that the final answer is true.

Question 3: If you were to cross a cream female to a white male, what ratio of body colours would you expect to see in their offspring? Explain. (10%)

Answer 3: Using the same notation as earlier:

$$\text{cream female}\left(C^w/C^b\right)\times\text{white male}\left(C^w/C^w\right)$$

$$=50\%\text{ cream}\left(C^w/C^b\right)+50\%\text{ white}\left(C^w/C^w\right)$$

Again, you may wish to use a Punnett square to prove this.

Question 4: If you were to cross a light brown female to a cream male, what ratio of body colours would you expect to see in their offspring? Explain. (10%)

Answer 4: Using the same notation as earlier:

$$\text{light brown female}\left(C^b / C^b\right) \times \text{cream male}\left(C^w / C^b\right) =$$

$$50\% \text{ cream}\left(C^w / C^b\right) + 50\% \text{ light brown}\left(C^b / C^b\right)$$

Again, you may wish to use a Punnett square to prove this.

Question 5: Supposing Mrs Cavy is correct in that the guinea pig body-colour gene you've identified encodes an enzyme in a pigment pathway, explain how variation at a single gene can generate three different body colour phenotypes? (20%)

Answer 5: Enzymes are proteins that catalyse the conversion of one molecule into another. In this case, as you are investigating coat colour, it is reasonable to think that an enzyme might catalyse the conversion of one pigment into another pigment (so effectively 'convert' one colour to another). As the alternate alleles at this gene are white and light brown, you can hypothesize that the enzyme encoded by this guinea pig body-colour gene converts a white pigment into a light-brown pigment.

Therefore the light-brown allele (C^b) is the wild-type allele that makes a functional enzyme, and the white allele (C^w) is the mutant allele that either doesn't make an enzyme, or makes one that doesn't work (this will depend on the precise nature of the mutation in the DNA and its consequence for transcription or translation). The full explanation of the genotypes and phenotypes would be:

C^b/C^b genotypes make a lot of enzyme that catalyses the conversion of all the white pigment to light-brown pigment, so the guinea pigs are light brown;

C^w/C^w genotypes make no enzyme, or make an enzyme that doesn't work, so cannot catalyse the conversion any white pigment, so the guinea pigs remain white;

C^w/C^b genotypes make some functional enzyme from their one wild-type allele, so can catalyse the conversion of some white pigment to light-brown pigment. Therefore they are a mixture of white and light-brown pigment and appear cream.

Note that the reverse, an enzyme catalysing the conversion of light-brown pigment to white pigment, is equally plausible. There are also possible answers where the gene encodes a protein that is not an enzyme (as you will discover in question 6).

Question 6: If the guinea pig body-colour gene does not encode an enzyme in a pigment pathway, speculate on what other kind of protein the body-colour gene might encode. In addition, explain how it might generate the three body colour phenotypes caused by variation at the body-colour gene. (20%)

Answer 6: Here you are asked to speculate; so as long as your answer is plausible and fits in with pattern of data then it is a good answer. It follows that there are many possible good answers, but it should look something like this.

If the gene does not encode an enzyme, then it must encode a different class of protein involved in pigmentation. Maybe light-brown pigment is made in specialist pigment-producing cells and carried to the hair. Perhaps the gene encodes a 'pigment-carrying protein'.

Therefore the light-brown allele (C^b) is the wild-type allele that makes a functional pigment-carrying protein, and the white allele (C^w) is the mutant allele that either doesn't make a pigment-carrying protein, or makes one that doesn't work. The full explanation of the genotypes and phenotypes would be:

C^b/C^b genotypes make a lot of pigment-carrying protein that carries all the light-brown pigment to the hairs, so the guinea pigs are light brown;

C^w/C^w genotypes make no pigment-carrying protein, or make a pigment carrying protein that doesn't work, so no light-brown pigment is carried to the hairs and the guinea pigs remain white;

C^w/C^b genotypes make some functional pigment-carrying protein from their one wild-type allele, so can carry some light-brown pigment to the hairs and appear cream.

Note that the reverse, a pigment-carrying protein carrying a white pigment to the hair, is equally plausible.

Again, there are other possible, plausible answers.

Chapter 1.9: The Mysterious Disappearance of Midnight

Question 1: If you want to retain the body of a recently killed animal for forensic analysis, why is putting it in a paper bag, rather than a plastic bag, a sensible idea? (10%)

Answer 1: If you wrap a dead animal in plastic it tends to 'sweat', encouraging bacteria and other organisms to grow. This accelerates the process of decomposition, therefore compromising the crime scene with 'foreign' DNA. This happens at a much slower rate in a paper bag.

Question 2: Why should the saliva sample give a better DNA profile than the DNA sample from the wound? (10%)

Answer 2: A saliva sample from the unbroken fur of the rabbit is less likely to be contaminated with DNA from the host (Midnight). A saliva sample from a wound will always be contaminated with host DNA.

Question 3: Briefly explain how PCR can be used to amplify STRs. (10%)

Answer 3: The size of an STR depends on the number of tandem repeats of a short DNA sequence. For example if the repetitive sequence is TGC then a 5-repeat sequence (TGCTGCTGCTGCTGC) is 15 nucleotides long whereas a 7-repeat sequence (TGCTGCTGCTGCTGCTGCTGC) is 21 nucleotides long. If you carry out PCR using primers that flank (sit just outside) the repeat, then the size of the PCR fragment produced will reflect the number of TGC repeats.

Question 4: How can STRs be used to identify a specific species? (10%)

Answer 4: Different species have different STRs at different positions on different chromosomes. As the sequences flanking an STR within a species are unique to that species and are found in every individual in that species, you can design a set of PCR primers that only amplify the STRs from a specific species. If the species-specific primers successfully amplify the targeted STRs, then the host animal must be from that species. If the species-specific primers amplify nothing, then the host animal must be from a different species.

Question 5: How can STRs be used to distinguish between specific individuals within a species? (5%)

Answer 5: Each individual within a species has the same STRs in precisely the same positions on the same chromosomes. Where they differ is in the number of repeats within each STR. Therefore, every individual within a species has the same number of amplified DNA fragments but the size of these fragments will be different.

Question 6: Using the DNA evidence in Fig. 1.9.3, explain which animal's saliva has been found on the body of Midnight? (5%)

Answer 6: Within Fig. 1.9.3, lane 3 matches lane S. The STR profile of Diablo, dog 3, is a perfect match to that of the saliva on Midnight.

Question 7: What statement should Dr Watson make to PC Holmes regarding which animal (or animals) attacked Midnight? (10%)

Answer 7: In a forensic analysis of this type, the STR profile simply places the owner of that saliva, in this case Diablo, at the crime scene. It would be for the police to establish why saliva matching Diablo's is found there.

Question 8: What would be the genetic relationship between two people if they shared all 26 STRs? (5%)

Answer 8: They must be monozygotic (identical) twins. This is because, as the name suggest, monozygotic twins originate as a single fertilized egg that divides in two very early in development to generate two genetically identical individuals. They could also be identical twins within a set of triplets or larger multiple birth.

Question 9: How many STRs do the two foxes (8 & 9) have in common? (5%)

Answer 9: They have one STR in common.

Question 10: What, if any, is the genetic relationship between the two foxes? (5%)

Answer 10: They are unrelated, or – because they have just one STR in common – they are at least not closely related.

Question 11: How many STRs do the three rottweiliers (1, 2 & 3) have in common? (5%)

Answer 11: Rottweillers 1 and 2 have 5 of their 8 STRs in common, 2 and 3 have 6 STRs in common, and 1 and 3 have 7 STRs in common.

Question 12: How many STRs do the three greyhounds (4, 5 & 7) have in common? (5%)

Answer 12: Greyhounds 4 and 5 have 7 STRs in common, 4 and 7 have 5 STRs in common and 5 and 7 have 6 STRs in common.

Question 13: Speculate on why the nature of STR variation in dogs is so different from that of foxes and humans. (15%)

Answer 13: There is much less STR variation within a breed of dogs than there is in foxes or humans, and this reflects the lower DNA variation in general within a breed of dog than within foxes or humans. This is because dog breeds are often highly inbred.

Acknowledgements

Thanks are due to Lucy Webster of the Wildlife DNA Forensic Unit based at the Science and Advice for Scottish Agriculture (SASA) laboratories near Edinburgh, in collaboration with the Royal Zoological Society of Scotland (RZSS) for help with this chapter.

Further reading

Dawnay, N., Ogden, R., Thorpe, R.S. *et al.* (2008) A forensic STR profiling system for the Eurasian badger: a framework for developing profiling systems for wildlife species. *Forensic Science International Genetics* **2**:47–53.

Chapter 1.10: RANCID

This chapter is based on the *tyrosinase* gene in the cat.

Question 1: Speculate on how the black/white body colour might be inherited. (10%)

Answer 1: As the F1 resemble the black parent and the F2 are in a 3:1 ratio of black to white, the simplest explanation is that this must be a single gene with two alleles where the black allele is dominant to the recessive white allele.

Question 2: If the body-colour gene encodes an enzyme, speculate on how it may cause the body colour phenotype. (10%)

Answer 2: The enzyme may convert a white pigment substrate into the black pigment. Wild-type (black) alleles would make a functional enzyme and convert all the white substrate to black pigment. The mutant (white) alleles make no enzyme, so do not convert any white substrate to black pigment, so remain white. Heterozygotes have a wild-type (black) and mutant (white) allele, where the dominant black allele makes a functional enzyme that converts the white substrate to a black pigment.

Question 3: Speculate on how the black/brown colour is inherited? (10%)

Answer 3: As the F1 resemble the black parent and the F2 are in a 3:1 ratio of black to brown, this must be a single gene with two alleles where the black allele is dominant to the recessive brown allele.

Question 4: What is the genetic basis of the three coat colours (black, brown and white)? (10%)

Answer 4: Taken together, these data suggest that the three coat colours are caused by different alleles at a single gene, where black is dominant to brown and both of them are dominant to white.

Question 5: In light of your answer to question 4, speculate on the type of mutations that may cause the phenotypes caused by the mutant alleles and explain their effect on the enzyme they encode. In addition, explain why you see the dominance/recessive relationships described earlier. (20%)

Answer 5: If this is a single gene with three possible alleles (black, brown and white), the black allele would be the wild-type functional copy that produces lots of functional enzyme for a step where a white substrate is converted to a black pigment. This will mask (be dominant to) any other allele. The white allele would be entirely non-functional and cannot convert any white substrate to black pigment, so the cat remains white. This could be due to a mutation in the control region of the gene that leads to no transcription, or a coding mutation such as a non-sense or missense mutation, where the latter must directly affect and stop protein/enzyme function This is likely to be recessive as the white allele has no function. The brown allele is best thought of as making some pigment but not at the normal rate. Therefore, the cats are not fully pigmented and appear intermediate and brown. The mutation here could be a control region mutation that decreases transcription, meaning the cats have less enzyme than normal. Alternatively, the brown allele may be caused by a missense that changes the quality of the enzyme, again meaning it makes less pigment.

Question 6: How might a mutation in a gene that encodes an enzyme lead to the phenotypes you see Meaow? (10%)

Answer 6: The colour pattern suggests that the enzyme is active in some parts of the cat and inactive in others. It's just about possible to imagine this could be a mutation in the control region of the gene, so it's only transcribed in the black-pigmented parts of the cat. However, given that the gene is normally expressed in all hairs (at least as far as their coat is concerned), it's difficult to see how a control mutation could generate a tissue-specific expression pattern. Note that the gene/protein only seems to be active in the periphery of the cat. So an alternative explanation is that the pattern is generated because of a mutation in the coding region of the gene that affects protein function. As the periphery of that cat is likely to be cooler than the core parts of the cat, this may be a temperature-sensitive mutation. In this case, a missense mutation affects the thermal stability of the protein; in warm parts of the cat the enzyme is inactive and the cat remains white, whereas in the cooler peripheries the enzyme is functional and these areas of the cat are black.

Question 7: If Meaow, or her Siamese brother, had kittens with Lucky, Chocolate or Snowy, what would the phenotype of each of the three sets of kittens be? (20%)

Answer 7: Taking each pair in turn:
black × Siamese would give all black;
white × Siamese would give Siamese;
brown × Siamese would give brown with Siamese peripheral black superimposed.

Question 8: Speculate on why Snowy, and other white-pigmented mammals, might be hearing impaired. (10%)

Answer 8: A key component of hearing are the tiny hairs within the ear. It may be that a lack of pigmentation within those hairs may affect their capacity to enable hearing.

Chapter 2.1: Otto's Finger

Question 1: As Otto was found in Europe, he must have died within the past 40 000 years, but how could you determine more precisely when he died? (5%)

Answer 1: The C14 method (radiocarbon dating) is the standard means of dating organic material. Tissues of all organisms absorb the C14 isotope from the atmosphere. When the host organism dies, the flow of C14 into the organism stops and the C14 isotopes within the host organism gradually disappear. This is because C14 has a half-life of 5730 years. As you know the level of C14 in the atmosphere and you can calculate the current level of C14 in the host using mass spectrometry, you can calculate the age of the host. In other words, organic materials lose C14 over time, so the lower the proportion of C14, the older the specimen.

Question 2: Why do studies of ancient remains often focus, at least initially, on mitochondrial DNA (which is precisely 16 569 bp long) rather than nuclear DNA (which is approximately 3 billion base pairs long)? (10%)

Answer 2: There are three reasons. Firstly, mtDNA is passed down the maternal line in an uninterrupted fashion and (as the mutation rate is constant) is attractive to researchers as there is no mitochondrial DNA recombination so mutations absolutely reflect maternal evolutionary history. Secondly, until fairly recently DNA sequencing technology was relatively 'primitive', so acquiring a full mtDNA profile (of 16 569 bp) was realistic, whereas attaining a full nuclear DNA profile from ancient remains was almost impossible. Spectacular recent improvements in DNA sequencing technologies mean that getting a full nuclear DNA profile is increasingly realistic. Thirdly, for reasons that are not entirely clear, nuclear DNA seems to decay at least twice as fast as mtDNA.

Question 3: From how many different people who were alive six generations ago have you acquired (a) your nuclear DNA and (b) your mitochondrial DNA? Are there any circumstances under which these theoretical estimates could be larger or smaller? (10%)

Answer 3: To take each question in turn:

In six generations your nuclear DNA must come from 64 people, 32 of whom are male and 32 of whom are female. This comes from the simple calculation that you have two parents (1 generation of ancestors), four grandparents (2 generations of ancestors), eight great-grandparents (3 generations ago), and so on. However, in small, inbred populations this cohort may be smaller. For example, cousins share recent ancestors (a grandparent) so the same ancestor may be represented more than once in the last 6 generations. Therefore, 64 is the highest number possible, and it may be smaller than that. It cannot be larger.

In six generations your mitochondrial DNA can only come down the maternal (female) lineage. Therefore you would have one source of your mitochondrial DNA.

Question 4: Use the data in Table 2.1.1 to construct an evolutionary tree showing the relationship between all 20 individuals. (20%)

Answer 4: The rationale here is to look for common nucleotide changes. For example, Otto, Oetzi, Adelina, Bruno and Hengist share the 16 362 change, so they must form a group. And so on. The entire relationship is shown in the following figure:

Nucleotide changes along a lineage are shown in black. Names of the 'owner' of each mitochondrial DNA sequence are shown in red. The blue names illustrate the subtypes of the K haplogroup (this information is not given in the original question). This tree can be drawn in many equally valid ways but as long as it is topologically correct, it is fine.

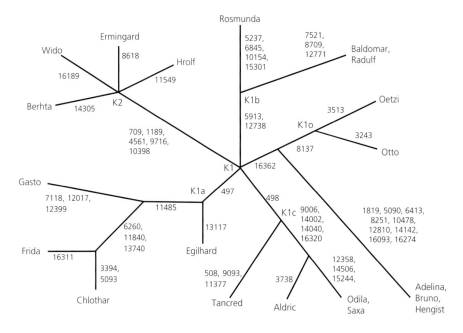

Question 5: Is the evolutionary tree you have constructed robust and reliable? How could you improve on its reliability? (5%)

Answer 5: It is robust, as each mutation is rare, and you have been told that each mutation has only occurred once in the lineage. Note that in reality several mutations at the same site are of course possible, and in a small dataset that can significantly confuse the data.

Question 6: What is the evidence that the MELAS family is suffering from a mitochondrial condition? (10%)

Answer 6: Mitochondrial conditions are inherited down the female line so every affected mother has the potential to pass the mutation (and therefore the condition) to all their sons and daughters as they inherit their cytoplasm (and therefore their mitochondria) from their mother. Affected fathers do not pass their cytoplasm (and therefore do not pass their mitochondria) to their children, so their children will not be affected unless the father's partner (the children's mother) is also affected. This is broadly true of this pedigree, though you might expect individuals II:6 and III:5 to be affected as the condition seems to 'skip' through them and similarly you would expect individuals III:9, IV:3, IV:4 and IV:5 to be affected as they have affected siblings. You have been told that the condition is variable in severity and age of onset, so it is possible that the individuals that you suspect are carrying the tRNA(Leu) A3243G mutation but who seem to be unaffected, may be too young to show symptoms.

Question 7: Why do mutations in the mitochondrial genome cause such variable phenotypes, and in some cases even seem to skip generations? (10%)

Answer 7: Mitochondrial conditions, at least those in which (like A3243G that causes MELAS) the causal mutation is within the mitochondrial genome, are very variable in severity, even within families. This is mainly a result of heteroplasmy. Unlike nuclear genes, where the frequency of a specific mutant allele within an individual can only be 0/2 (wild-type), 1/2 (heterozygous) or 2/2 (homozygous mutant), each cell has thousands of mitochondria and millions of copies of mitochondrial DNA. Therefore, a mitochondrial mutation can theoretically comprise anything from 0% to 100% of the mitochondrial DNA. Of course, if 100% of the mtDNA is mutant then the host may be severely affected and may even die. By definition, if 100% of your DNA is wild-type you will have no mitochondrial disease phenotype. For most mitochondrial conditions there is an approximate 'threshold' effect, where individuals with (say) 60% mutant mtDNA or higher are affected, and those below 60% are not. Above the threshold, the higher the proportion of mtDNA that is mutant, the more severe the condition and perhaps the earlier the age on onset. This explains the variation in severity of phenotype. The relative frequency of mutation versus wild-type mtDNA can change significantly between generations, which explains the differences seen between generations, including the fact that mitochondrial conditions can seem to 'skip' generations.

Question 8: Speculate on the functional consequence of a mutation in a mitochondrial tRNA(Leu) gene? Why should it lead to such a broad range of phenotypes affecting so many organs and systems in the body? (10%)

Answer 8: tRNA genes are an important component of the translational machinery and they play a key role in the accuracy of translation. The anticodon on the tRNA matches the codon on the mRNA and this allows the correct amino acid to be incorporated into the correct position on the growing peptide/protein. A mutation in a tRNA(Leu) gene could prevent leucine being incorporated into the a growing peptide/protein, or could mean it is incorporated into the wrong position in the growing peptide/protein. As a consequence, the translational machinery will be less accurate and a proportion of the protein produced will be synthesized incorrectly and will have the wrong amino acid sequence. As this has the potential to affect every protein type made in the mitochondria, and all cells are likely to be affected, leading to a broad range of mitochondrial dysfunction phenotypes.

Question 9: Propose a mechanism by which the C-13910T and G-22018A mutations in the human *LCT* gene might lead to a change in human phenotype from lactose intolerant to lactose tolerant. (10%)

Answer 9: The difference in phenotype between lactose intolerant and lactose tolerant individuals is that the *LCT* gene is actively transcribed only in early childhood (intolerant) or throughout the entire lifespan (tolerant). Therefore, the two SNPs (mutations) must affect gene expression. As the two SNPs are thousands of bases upstream of the *LCT* gene, they must affect transcription and presumably must prevent the *LCT* gene from being switched off. The mutations may prevent binding of an age-induced transcriptional repressor.

Question 10: Otto is homozygous for the lactose intolerant allele, but some of his friends may well have been heterozygous. Would Otto's heterozygous friends have been lactose tolerant or intolerant or somewhat intermediate? Explain why. (10%)

Answer 10: Otto's heterozygous friends would probably have been lactose tolerant as they have one allele being expressed during childhood and the other expressed throughout their entire life. The phenotype would therefore reflect that of a dominant gain-of-function, 'expressed throughout their life' lactose-tolerant allele. However, it is possible that having one *LCT* gene expressed throughout their life would actually only give partial tolerance.

Chapter 2.2: The Mystery of Muckle Morag

This chapter is based on a 90-minute exam paper taken by the Level 3 Genetics class at the University of Glasgow in May 2013.

Question 1: Discuss the evidence supporting the idea that the 'Cameron' nose is caused by an X-linked recessive condition? (5%)

Answer 1: There is no male-to-male (father to son) transmission and it skips generations through females (grandfather to grandson transmission via an unaffected daughter/mother). All of these modes of inheritance are typical of an X-linked recessive condition.

As this is an inbred population, individuals coming into the family may also have the condition (such as woman III/9) or be carriers. This complicates the genetics. In a large outbred population you wouldn't expect so many (or indeed any) of the partners of the affected family to have the same rare genetic condition. The only exception would be where individuals affected by the same genetic condition are, for whatever reason, attracted to each other.

Question 2: Redraw the family tree showing, where possible and using appropriate notation, the *cameron* genotypes of all the members of the family. (10%)

Answer 2: The *cameron* genotypes are shown in the figure, where '+' is a wild-type *cam* allele, '−' is a mutant *cam* allele and '?' is an unknown *cam* allele:

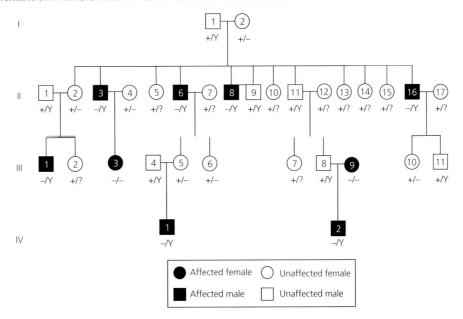

When a mutant allele (such as *cameron*) is recessive it is impossible to determine whether an unaffected individual is homozygous wild-type (+/+) or a heterozygous carrier (+/−) from their phenotype alone. However, their status can often be determined by looking at the phenotypes of their parents and/or children. Clearly, in an X-linked condition the genotype of each male can be determined because the genotype (+/Y or −/Y) absolutely reflects the phenotype (unaffected or affected)

Question 3: What is the evidence that *cameron* and the CAT STR are genetically linked? (10%)

Answer 3: Of the six affected males from within the family (II:3, II:6, II:8, III:1 and IV:1), five have inherited the mutant *cameron* allele along with the STR 9 allele. In addition, both of the affected women from within the pedigree (II:3 and III:9) have co-inherited the *cameron* allele and the STR 9 allele from an affected father. This is reinforced by the fact that STR 15 does not co-segregate with the condition. Note that IV:2 doesn't count in this respect because he inherits his mutant *cameron* allele from his mother, who is not descended from the original couple (I:1 and I:2)

Question 4: Calculate the genetic map distance between *cameron* and the CAT STR. (20%)

Answer 4: The figure shown here explains everything (although the question doesn't actually ask you to draw a figure). There are 8 parentals and 1 recombinant. So the map distance is (1/9) × 100 = 11.1.

Not every individual is informative. For an X-linked condition, recombination can only occur in a mother and you can only see whether recombination has occurred if the mother is heterozygous for both the condition (+/−) and the STR. You can then look for evidence of recombination in her sons and (occasionally) in her daughters. This excludes several individuals; for example, female III:5, who is heterozygous herself but has a homozygous mother, so she is uninformative, but you can detect whether or not recombination has occurred in her son.

In the figure, the affected *cameron* allele is shown linked to the red CAT STR allele.

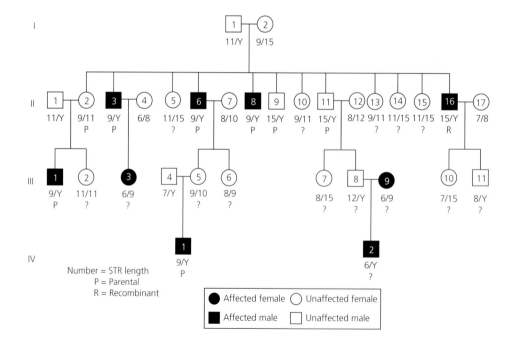

Question 5: If woman II:2 has a child with her younger brother II:11, what is the probability that they will have a Cameron-nosed son of CAT STR 11/Y? (5%)

Answer 5: There are three frequencies that you need to multiply together. Son (0.5) × Cameron-nose allele (0.5) × recombination (0.11) = 0.0275. Therefore, the answer in percentage terms is 2.75%.

Question 6: Explain how you would use the 12 whole genome sequences, the family tree, the mapping studies and your knowledge of human mutations to identify the mutant DNA sequence of the *cameron* gene. What problems might you encounter? (20%)

Answer 6: The key point is that the mutants must have the same mutation (specific DNA change) because all the affected individuals are related. So if you line up six affected sequences and six unaffected sequences, the mutation must be present in the DNA sequence from the predicted region in all six affected individuals, but in none of the six unaffected individuals. However, since the family is inbred and the mutation is identical by descent over fairly few generations, the haplotype, meaning the continuous identical DNA sequence that all affected individuals and all unaffected individuals have in common, may be quite large because there hasn't been a lot of opportunity for recombination to occur. So, if the affected individuals differ in DNA sequence from the unaffected individuals in several ways, you might first look for changes that are likely to cause a profound phenotypic change. Something like a STOP codon or deletion/insertion affecting a gene would be fortuitous. A missense codon may or may not affect a phenotype depending on the specific amino acid change, especially whether the two amino acids are similar and whether the amino acid is in a key functional domain of the protein. From sequence data alone it may be very difficult to determine whether a mutation outside the coding region was the cause of the mutant phenotype, unless you can demonstrate that the mutation affects splicing, or alternatively affects the binding of transcription factors which change the quantity of mRNA produced by the gene. What you are looking for here is an understanding of the strategy, especially the problems you are likely to encounter. One way of resolving this would be to collect more sequence data from more affected and unaffected individuals.

Question 7: In light of your preliminary analysis, speculate on how the recessive missense mutation might cause the *cameron* mutant phenotype. (10%)

Answer 7: In the mutants, the receptor protein is in the cytoplasm and not in the cell membrane. This could be because the mutant Cameron protein cannot be delivered or cannot find its way to the cell membrane, or because it has a conformational change such that even if it is delivered to the membrane, it cannot integrate into it. The answer needs to explain why the mutation is recessive. The best way to accommodate this into your answer is to ensure that the mutation results in a simple loss of protein function, as in the earlier examples. Presumably, the missense mutation in the mutant protein means that the (altered) amino acid is directly involved in either the transport of the CAM protein to the membrane, or it's insertion into it. Critically, you are looking for any idea that is plausible.

Question 8: Speculate on how a dominant missense mutation in the *cameron* gene of the Hooter family might result in large noses. (10%)

Answer 8: As this is a dominant missense mutation, and the Hooter phenotype of big noses is the opposite of the small misshapen Cameron nose, a reasonable model would suggest that in the Hooter family, the cell-surface, membrane-bound receptor encoded by the *cameron* gene is more active so it sends more signal. There are several possibilities. The receptor could be constitutively active, signalling in the absence of its ligand. Alternatively, the receptor might normally interact with a repressor molecule and the mutation stops this from happening, so the mutant receptor is irrepressible. Or the receptor protein could be hyperstable, so it behaves and signals normally, but simply survives longer. Note that the answer has to be an effect on protein function, as mRNA and protein levels are apparently normal.

Question 9: How would you test the model you proposed speculating on how the dominant missense mutation in the *cameron* gene of the Hooter family causes the large nose phenotype? (10%)

Answer 9: Clearly, the answer must focus on the model proposed in the answer to question 8, possibly looking at receptor function, including binding/ligand assays. Again, anything plausible is potentially a good answer. The key here is to have a clear experimental aim and appropriate controls.

Chapter 2.3: *Drosophila hogwashii*

Question 1: What affect does EMS have on DNA? (5%)

Answer 1: EMS is a base-modifier mutagen that is an alkylating agent. It transfers ethyl groups to bases, resulting in altered base pairing. Specifically, it converts guanine (G) into 7-ethyl guanine (7G). As 7G pairs with thymine (T) rather than cytosine (C) the consequence of EMS exposure is to change a G:C pair into A:T, a process called transition. These random single base-pair changes, or point mutations, may or may not have a consequence for the phenotype, which depends entirely on where the G:C to A:T change occurs in the genome.

Question 2: How could the students determine whether the *vol-au-vent* mutation is dominant or recessive? (5%)

Answer 2: Simply cross a mutant fly to a wild-type fly and look at the phenotype of the F1 offspring, which must be heterozygous (+/−). If the F1 heterozygous flies have a wild-type (green-eye) phenotype then the mutation is recessive, whereas if the F1 heterozygous flies have a mutant (yellow-eye) phenotype the mutation is dominant.

   Things will be more complicated if the mutation is on one of the sex-chromosomes (the X-chromosome in *Drosophila hogwashii*, for example). In this case you can only judge dominance or recessivity in females, as they are the sex that has two X-chromosomes.

Question 3: How could the students determine whether the *vol-au-vent* mutation is X-linked or on chromosome 2 or 3 or 4? (10%)

Answer 3: X-linked genes (alleles) cannot be passed from father to son. To show this, set up a cross between a *vol-au-vent* homozygous mutant female and a homozygous wild-type male. If the mutation is X-linked, all F1 males will be *vol-au-vent* mutants; however, if the mutation is autosomal, none of the F1 males will be *vol-au-vent* mutants. If the mutation is not on the X-chromosome then it could be on any of chromosomes 2 or 3 or 4. The easiest way to determine this is to see whether the mutation is co-inherited with other mutant alleles. It can only be co-inherited and therefore show linkage with other genes/mutations that are on one chromosome (either 2 or 3 or 4).

Question 4: Using appropriate notation and showing full genotypes, draw the students' crossing scheme. (10%)

Answer 4: You should draw something along the following lines:

$$\frac{vol^- \ hag^+ \ sho^+}{vol^- \ hag^+ \ sho^+} \quad \times \quad \frac{vol^+ \ hag^- \ sho^-}{vol^+ \ hag^- \ sho^-}$$

(Female *vol-au-vent* × male *haggard shortbottom*)

$$\frac{vol^- \ hag^+ \ sho^+}{vol^+ \ hag^- \ sho^-}$$

F1 all wild-type as here (and heterozygous at each gene)

Test cross $\quad \dfrac{vol^- \ hag^+ \ sho^+}{vol^+ \ hag^- \ sho^-} \quad \times \quad \dfrac{vol^- \ hag^- \ sho^-}{vol^- \ hag^- \ sho^-}$

(F1 female triply heterozygous × male triple mutant)

There are eight possible genotypes of offspring from the test cross. All eight have one chromosome *vol⁻ hag⁻ sho⁻* from their father, so only the other chromosome is shown here.

vol⁺	*hag⁺*	*sho⁺*
vol⁻	*hag⁻*	*sho⁻*
vol⁺	*hag⁺*	*sho⁻*
vol⁻	*hag⁻*	*sho⁺*
vol⁺	*hag⁻*	*sho⁺*
vol⁻	*hag⁺*	*sho⁻*
vol⁻	*hag⁺*	*sho⁺*
vol⁺	*hag⁻*	*sho⁻*

Question 5: Use the data in Table 2.3.1 to draw a map clearly showing the position of the *vol-au-vent* mutation relative to *haggard* and *shortbottom* genes. (15%)

Answer 5: Map distances are:
vol to *hag* ((35 + 29 + 2 + 3)/1000) × 100 = 6.9
vol to *sho* ((2 + 3 + 61 + 65)/1000) × 100 = 13.1
hag to *sho* ((35 + 29 + 61 + 65)/1000) × 100 = 19.0

Question 6: Why don't the map distances add up precisely? (5%)

Answer 6: Map distances never add up precisely as double crossovers ensure that large distances are underestimated. This is because when you only look at recombinants between two genes (A and B below) on a chromosome, any double crossovers between the two genes will result in the original allelic pairings within a chromosome (A–B and a–b in the image shown here) being maintained in their original combinations, even though crossing-over has occurred twice between them.

In addition, it seems that a crossover event actively precludes another crossover event very close by, a phenomenon called 'interference'. So double crossovers are possible and do occur, but at a lower frequency than you might expect.

Question 7: Redraw the deletion map from Fig. 2.3.4 and show approximately where the *vol-au-vent* mutation lies. (10%)

Answer 7: The *vol-au-vent* mutation must be where leo and vipera are deleted (white), but meles and aquila are not (black). This is indicated on the image here and is the area between the two vertical lines.

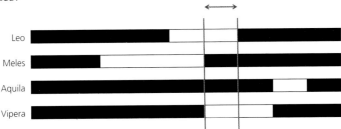

Question 8: Draw a crossing scheme for the deletion mapping, clearly showing the frequency of every genotype and phenotype in each generation when (a) you start with a homozygous *vol-au-vent* female and cross her to a male carrying the leo deletion, and (b) you start with a homozygous *vol-au-vent* female and cross her to a male carrying the aquila deletion. (10%)

Answer 8: Taking both crossing-schemes in turn:

(a) Cross female vol^-/vol^- mutant with male leo^-/leo^+ (carrying the deletion in a heterozygous state) to give two classes of offspring vol^-/leo^- (mutant phenotype) and vol^-/leo^+ (wild-type phenotype). The mutation is in the deletion so the cross concerns three different alleles: leo^+ (which is the same as vol^+), vol^-, leo^-, at one gene. The critical issue is that deletion mutations are recessive lethal (this is confirmed in the question), so the deletion strain you start with must be heterozygous.

$$\frac{vol^-}{vol^-} \times \frac{leo^-}{leo^+}$$

Female *vol⁻* mutant crossed to male heterozygous for the leo deletion

$$\frac{vol^-}{leo^-} \times \frac{vol^-}{leo^+}$$

F1: equal numbers of yellow eye (*vol⁻/leo⁻*) and green eye (*vol⁻/leo⁺*) flies

(b) Cross female vol^- aquila$^+$ / vol^- aquila$^+$ (homozygous mutant) with male vol^+ aquila$^-$ / vol^+ aquila$^+$ (which again is heterozygous for the deletion) to give two classes of offspring vol^- aquila$^+$ / vol^+ aquila$^-$ (wild-type phenotype and heterozygous for both) and vol^- aquila$^+$ / vol^+ aquila$^+$ (wild-type phenotype and heterozygous for *vol* only). This time the deletion and mutation do not overlap so treat them as two different genes. Again the critical issue is that deletion mutations are lethal (this is confirmed in the text), so the male you start with must be heterozygous.

$$\frac{vol^- \; aquila^+}{vol^- \; aquila^+} \times \frac{vol^+ \; aquila^-}{vol^+ \; aquila^+}$$

Female *vol⁻* mutant crossed to male heterozygous for the eagle deletion

$$\frac{vol^- \; aquila^+}{vol^+ \; aquila^-} \quad \frac{vol^- \; aquila^+}{vol^+ \; aquila^+}$$

F1: all flies have green eye colour, as all are heterozygous for *vol* (*vol⁺/vol⁻*)

Question 9: Briefly describe how the students could prove that the pigment enzyme gene is *vol-au-vent*? (5%)

Answer 9: The best way is gene rescue. Create a transgene containing a wild-type copy of the pigment pathway gene and cross it into the *vol-au-vent* mutant strain. If a *vol-au-vent* mutant carrying a wild-type transgenic copy of the pigment pathway gene has wild-type eye colour (demonstrating gene rescue), then it's the right gene. Only a wild-type copy of the correct transgene can do this. Other possibilities include RNAi, although there are two potential issues with RNAi. Firstly, RNAi knockdown varies widely in its efficacy so negative results are meaningless. In addition, RNAi may hit a different gene, so positive results may be meaningless too. Therefore an RNAi knockdown approach is not as good as gene rescue. Sequencing the host pigment pathway gene provides some supportive evidence, especially if you discover a nonsense or frame-shift mutation which you might expect to cause a phenotype, but it actually proves nothing.

Question 10: What are the relative advantages and disadvantages of EMS mutagenesis as against *P*-element mutagenesis? (5%)

Answer 10: EMS mutagenesis causes point mutations that are effectively random, so any gene can be mutated. *P*-elements tend to mutate some genes at very high frequencies and some not at all.

EMS mutagenesis causes point mutations, which can have a wide variety of consequences for control and coding regions, so a wide variety of mutations, some subtle, some catastrophic and many in-between, can be created. *P*-elements are insertions so normally knock-out a gene causing null mutations.

EMS is very efficient as it creates lots of new mutations within a fly. *P*-elements create many fewer mutations within each fly.

When a *P*-element has inserted into a gene you can find that host gene simply by looking for the DNA sequence adjacent to the known *P*-element sequence, so it's relatively easy to find the mutated gene. It is much harder to find the EMS-mutated gene as it simply has a single base-pair change. Most of the above are broadly true of any transposon vs chemical/radiation mutagenesis scheme in any organism.

Question 11: Speculate on what eye-colour the *vol-au-vent*; *malformed* double mutant would have if Hermes' theory is correct? (10%)

Answer 11: If they are in the same pathway then the simplest explanation is that each gene encodes a different enzyme on that pathway. It follows that there are two possible answers, both of which are equally likely (and therefore equally correct):

So a double mutant by this scheme has either blue or yellow eyes (but not both), depending on the order of genes/enzymes along the linear pathway. The gene early in the pathway is epistatic to the gene later in the pathway.

Question 12: In fact the *vol-au-vent*; *malformed* double mutant flies have white-coloured eyes. Speculate on why. (10%)

Answer 12: The white colour suggests that the flies have no pigment, so each gene must enable pigment synthesis independently. In other words:

In a *malformed* mutant you don't make yellow, so the eyes are blue.
In a *vol-au-vent* mutant you don't make blue, so the eyes are yellow.
In a double mutant you don't make anything, so the eyes are white.
In wild-type you make yellow and blue, so the eyes are green.

Chapter 2.4: The Curse of Lilyrot

Question 1: From a genetic perspective, what is the significance of the strains being described as 'true-breeding' and 'independently derived'? (5%)

Answer 1: 'True-breeding' plants are those that are homozygous for the characteristics you are studying. They are generally inbred. 'Independently derived' suggests that the three strains are only very distantly related, so you can be fairly confident that they carry different mutations, although this could include different mutations in the same gene.

Question 2: Given that you know the organisation of the biochemical pathways that generate lily flower colour, what is the simplest genetic explanation of the three F1 lily flower colours described in Table 2.4.1? (10%)

Answer 2: The easiest way to think about this us that you have three enzymes each of which is encoded by a gene. Whilst any reasonable notation will suffice, let's say that enzyme R is encoded by gene R, in addition enzyme Y is encoded by gene Y and enzyme B is encoded by gene B. The modified Fig. 2.4.1 here should clarify this:

'RED' pathway

$$\text{white} \xrightarrow{\text{Enzyme R}} \text{red}$$

'BLUE' pathway

$$\text{white} \xrightarrow{\text{Enzyme Y}} \text{yellow} \xrightarrow{\text{Enzyme B}} \text{blue}$$

Note that there are actually only two pathways, not three, and that blue pigment can only be made if yellow pigment is made first. You can say that the gene encoding the enzyme (Enzyme Y) catalysing the conversion of white pigment to yellow pigment is epistatic to the gene that encodes the enzyme (Enzyme B) that catalyses the conversion of yellow pigment to blue pigment.

Using standard notation, you can say that R^+, Y^+ and B^+ are the dominant, wild-type, enzyme-encoding alleles, whilst R^-, Y^- and B^- are the recessive, mutant, non-functional alleles. As the three original strains are true breeding, the three true-breeding strains must have the following genotypes:

BLUE:	$R^- R^-$	$Y^+ Y^+$	$B^+ B^+$
RED:	$R^+ R^+$	$Y^- Y^-$	$B^- B^-$
YELLOW:	$R^- R^-$	$Y^+ Y^+$	$B^- B^-$

The F1 plants derived from the three original true–breeding strains must therefore have the genotypes:

BLUE × RED:	$R^+ R^-$	$Y^+ Y^-$	$B^+ B^-$
BLUE × YELLOW:	$R^- R^-$	$Y^+ Y^+$	$B^+ B^-$
RED × YELLOW:	$R^+ R^-$	$Y^+ Y^-$	$B^- B^-$

Question 3: Regarding Table 2.4.1, if you were to cross two F1 blue flowering lilies together, what would be the ratio of flower colours in their offspring? (5%)

Answer 3: The F1 blue strain is derived from a cross between the true-breeding yellow strain, which can only make the yellow pigment (R^-R^- Y^+Y^+ B^-B^-) and the true-breeding blue strain which must be able to make yellow and convert it to blue (R^-R^- Y^+Y^+ B^+B^+). Therefore, you can ignore the *red* gene, because every lily in this cross is mutant for it (R^-R^-), and ignore the *yellow* gene as every lily in this cross is wild-type for it (Y^+Y^+) and can always make yellow pigment. The only question is therefore whether they convert yellow to blue.

Therefore, the F1 blue individuals have the genotype R^-R^- Y^+Y^+ B^+B^-.

Looking at this in a Punnett square (ignoring the R and Y genes, and remembering that every lily will therefore at least be yellow) you get:

	B^+	B^-
B^+	B^+B^+ blue	B^+B^- blue
B^-	B^+B^- blue	B^-B^- yellow

This is a standard 3:1 ratio.

Question 4: Regarding Table 2.4.1, if you were to cross two F1 orange flowering lilies together, what would be the ratio of flower colours in their offspring? (5%)

Answer 4: The F1 orange strain is derived from a cross between the original true-breeding red strain that must be homozygous wild-type for the ability to make red pigment in the flowers (enzyme R encoded by gene R), but mutant for both genes in the blue pathway (genes Y and B), and the original true-breeding yellow strain that must be homozygous wild-type for the ability to make the yellow pigment (enzyme Y encoded be gene Y), but mutant for both the *red* (R) and *blue* (B) genes. As both starting strains are mutant for the *blue* gene, you can ignore blue in the answer as every lily in this cross is homozygous mutant for the *blue* gene (B^-B^-).

Therefore the F1 orange individuals have the genotype R^+R^- Y^+Y^- B^-B^-.

Looking at this in a Punnett square (and ignoring the B gene) you get:

	R^+Y^+	R^+Y^-	R^-Y^+	R^-Y^-
R^+Y^+	R^+R^+ Y^+Y^+ orange	R^+R^+ Y^+Y^- orange	R^+R^- Y^+Y^+ orange	R^+R^- Y^+Y^- orange
R^+Y^-	R^+R^+ Y^+Y^- orange	R^+R^+ Y^-Y^- red	R^+R^- Y^+Y^- orange	R^+R^- Y^-Y^- red
R^-Y^+	R^+R^- Y^+Y^+ orange	R^+R^- Y^+Y^- orange	R^-R^- Y^+Y^+ yellow	R^-R^- Y^+Y^- yellow
R^-Y^-	R^+R^- Y^+Y^- orange	R^+R^- Y^-Y^- red	R^-R^- Y^+Y^- yellow	R^-R^- Y^-Y^- white

This is a standard 9:3:3:1 ratio.

Question 5: Regarding Table 2.4.1, if you were to cross two F1 purple flowering lilies together, what would be the ratio of flower colours in their offspring? (15%)

Answer 5: The F1 purple strain is derived from a cross between the true-breeding blue strain, which must be able to make the yellow pigment and convert it to blue ($R^-R^-\ Y^+Y^+\ B^+B^+$) and the true-breeding red strain which must be able to make the red pigment ($R^+R^+\ Y^-Y^-\ B^-B^-$).

Therefore the F1 are triple heterozygotes with the genotype $R^+R^-\ Y^+Y^-\ B^+B^-$.

Looking at this on a Punnett square you get:

	$R^+Y^+B^+$	$R^+Y^+B^-$	$R^+Y^-B^+$	$R^+Y^-B^-$	$R^-Y^+B^+$	$R^-Y^+B^-$	$R^-Y^-B^+$	$R^-Y^-B^-$
$R^+Y^+B^+$	R^+R^+ Y^+Y^+ B^+B^+ Purple	R^+R^+ Y^+Y^+ B^+B^- Purple	R^+R^+ Y^+Y^- B^+B^+ Purple	R^+R^+ Y^+Y^- B^+B^- Purple	R^+R^- Y^+Y^+ B^+B^+ Purple	R^+R^- Y^+Y^+ B^+B^- Purple	R^+R^- Y^+Y^- B^+B^+ Purple	R^+R^- Y^+Y^- B^+B^- Purple
$R^+Y^+B^-$	R^+R^+ Y^+Y^+ B^+B^- Purple	R^+R^+ Y^+Y^+ B^-B^- Orange	R^+R^+ Y^+Y^- B^+B^- Purple	R^+R^+ Y^+Y^- B^-B^- Orange	R^+R^- Y^+Y^+ B^+B^- Purple	R^+R^- Y^+Y^+ B^-B^- Orange	R^+R^- Y^+Y^- B^+B^- Purple	R^+R^- Y^+Y^- B^-B^- Orange
$R^+Y^-B^+$	R^+R^+ Y^+Y^- B^+B^+ Purple	R^+R^+ Y^+Y^- B^+B^- Purple	R^+R^+ Y^-Y^- B^+B^+ Red*	R^+R^+ Y^-Y^- B^+B^- Red*	R^+R^- Y^+Y^- B^+B^+ Purple	R^+R^- Y^+Y^- B^+B^- Purple	R^+R^- Y^-Y^- B^+B^+ Red*	R^+R^- Y^-Y^- B^+B^- Red*
$R^+Y^-B^-$	R^+R^+ Y^+Y^- B^+B^- Purple	R^+R^+ Y^+Y^- B^-B^- Orange	R^+R^+ Y^-Y^- B^+B^- Red*	R^+R^+ Y^-Y^- B^-B^- Red	R^+R^- Y^+Y^- B^+B^- Purple	R^+R^- Y^+Y^- B^-B^- Orange	R^+R^- Y^-Y^- B^+B^- Red*	R^+R^- Y^-Y^- B^-B^- Red
$R^-Y^+B^+$	R^+R^- Y^+Y^+ B^+B^+ Purple	R^+R^- Y^+Y^+ B^+B^- Purple	R^+R^- Y^+Y^- B^+B^+ Purple	R^+R^- Y^+Y^- B^+B^- Purple	R^-R^- Y^+Y^+ B^+B^+ Blue	R^-R^- Y^+Y^+ B^+B^- Blue	R^-R^- Y^+Y^- B^+B^+ Blue	R^-R^- Y^+Y^- B^+B^- Blue
$R^-Y^+B^-$	R^+R^- Y^+Y^+ B^+B^- Purple	R^+R^- Y^+Y^+ B^-B^- Orange	R^+R^- Y^+Y^- B^+B^- Purple	R^+R^- Y^+Y^- B^-B^- Orange	R^-R^- Y^+Y^+ B^+B^- Blue	R^-R^- Y^+Y^+ B^-B^- Yellow	R^-R^- Y^+Y^- B^+B^- Blue	R^-R^- Y^+Y^- B^-B^- Yellow
$R^-Y^-B^+$	R^+R^- Y^+Y^- B^+B^+ Purple	R^+R^- Y^+Y^- B^+B^- Purple	R^+R^- Y^-Y^- B^+B^+ Red*	R^+R^- Y^-Y^- B^+B^- Red*	R^-R^- Y^+Y^- B^+B^+ Blue	R^-R^- Y^+Y^- B^+B^- Blue	R^-R^- Y^-Y^- B^+B^+ White*	R^-R^- Y^-Y^- B^+B^- White*
$R^-Y^-B^-$	R^+R^- Y^+Y^- B^+B^- Purple	R^+R^- Y^+Y^- B^-B^- Orange	R^+R^- Y^-Y^- B^+B^- Red*	R^+R^- Y^-Y^- B^-B^- Red	R^-R^- Y^+Y^- B^+B^- Blue	R^-R^- Y^-Y^- B^-B^- Yellow	R^-R^- Y^-Y^- B^+B^- White*	R^-R^- Y^-Y^- B^-B^- White

Generally three independent genes give a ratio of 27:9:9:9:3:3:3:1 to give 64 in total.

In this case you get 27 purple, 12 red (including 9 red* above), 9 orange, 9 blue, 4 white (including 3 white* above) and 3 yellow, which again is 64. The 12 starred classes are due to epistasis, where at least one B^+ allele is present but the individual is Y^-Y^-, so there is no yellow pigment for the B enzyme to work on.

Question 6: Using appropriate notation, draw the full crossing scheme for the backcross undertaken by your aunt. (5%)

Answer 6: In this crossing scheme the two alleles for the *lilyrot* gene are the dominant sensitive allele L^s and the recessive resistant allele L^r. The two alleles at the *savage* leaf shape gene are the dominant wild-type S^+ and the recessive *savage* mutant S^-. Again, any sensible alternative notation is entirely reasonable.

In the cross the parental strains are:

lilyrot-sensitive with wild-type leaves ($L^s S^+ / L^s S^+$)

lilyrot-resistant with savage leaves ($L^r S^- / L^r S^-$)

The F1 lilies must all have wild-type leaves and be lilyrot sensitive with the genotype: ($L^s S^+ / L^r S^-$)

You then perform a backcross between the F1 wild-type leaved and lilyrot-sensitive lilies ($L^s S^+ / L^r S^-$) generated earlier to lilies that are homozygous for the two recessive mutations, namely a lilyrot-resistant with savage leaves strain ($L^r S^- / L^r S^-$)

This will generate four possible offspring:

Two classes with parental phenotypes:

lilyrot-sensitive with wild-type leaves ($L^s S^+ / L^r S^-$)

lilyrot-resistant with savage leaves ($L^r S^- / L^r S^-$)

Two classes with recombinant phenotypes:

lilyrot-sensitive with savage leaves ($L^s S^- / L^r S^-$)

lilyrot-resistant with wild–type leaves ($L^r S^{-+}/ L^r S^-$)

Question 7: Use a chi-squared test to determine whether the leaf shape and *lilyrot* genes are on the same or different chromosomes. (10%). Note that for one degree of freedom chi squared scores above 3.84 are considered significant ($p < 0.05$).

Answer 7: The null hypothesis is that the genes are on different chromosomes, in which case you would expect equal numbers of each of the four classes of offspring.

Phenotype	Observed number	Expected number	(O – E)	(O – E)²	$\dfrac{(O - E)^2}{E}$
wild-type sensitive	320	250	70	4900	19.6
wild-type resistant	190	250	60	3600	14.4
savage **sensitive**	210	250	40	1600	6.4
savage **resistant**	280	250	20	400	3.6
total	1000	1000			44.0

The chi-squared value is 44.0. Scores above 3.84 are statistically significant. Therefore the hypothesis that the genes are unlinked is incorrect. Therefore the genes must be linked on the same chromosome.

Question 8: Calculate the map distance between the leaf shape and *lilyrot* genes (5%)

Answer 8: Map distances are calculated using the formula:

$$(\text{recombinants}/\text{total}) \times 100$$

Therefore for this cross the map distance is:

$$((210 + 190)/1000) \times 100 = 40$$

Question 9: Describe a series of experiments that would allow you to identify the DNA sequence of the mutant gene which determines lilyrot resistance. How would you prove that this was the gene that, when mutant, caused the lilyrot-resistant phenotype? (10%)

Answer 9: Having calculated the approximate map position you might look for a candidate gene in the identified region that has an obvious mutation in it. Obvious mutations in these candidate genes include anything that disrupts the coding sequence (deletion, inversion, insertion) or a nonsense mutation. These nearly always cause a null mutant phenotype. Depending on the specific amino acid change and the position of that amino acid in the protein, a missense mutation can have anything between a catastrophic and a neutral effect on a protein encoded, so it is harder to say definitively that a missense mutation is causal. Non-coding mutations that disrupt upstream control sequences or splicing sequences could cause the phenotype, but it's harder to prove unequivocally that these mutations actually cause the phenotype. All of this is much harder if there are many mutations in many candidate genes.

As the identified region contains lots of genes, further recombination or gene mapping might help to narrow the region down. This coupled with comparative DNA sequence and/or microarray and/or RNA-seq analysis of this region may identify the mutant gene (or at least narrow the field). But you must do DNA sequencing, microarrays or RNA-seq in a wild-type strain and a mutant strain and compare the sequence/expression data between them.

The best way of proving that a specific mutation is the cause of a mutant phenotype is gene rescue, where a wild-type transgene is introduced to the mutant lily. If it rescues (restores) the wild-type phenotype the transgene must be the same gene as the mutant gene. This will only work if the mutation is recessive.

Question 10: How might a four base-pair deletion in the coding region of a cell-surface membrane-bound receptor gene lead to lilyrot resistance? How might you test your theory? (10%)

Answer 10: Viruses infect cells by binding to one or more molecules on the cell's surface and injecting their viral DNA (or RNA) into the cell. This cell-surface membrane-bound receptor may be the receptor that the lilyrot virus binds to, thus enabling infection. The four base-pair deletion within the coding region would cause a frame-shift mutation and make a protein that is unlikely to bind to the lilyrot virus. To test this you could undertake some kind of binding assay to determine whether the lilyrot virus can bind to the wild-type and mutant cell-surface membrane-bound receptors. You could also generate other mutations in the gene and see whether these also make lilies resistant to lilyrot.

Question 11: For each of the four mutations listed in Table 2.4.3, speculate on whether they might cause a lilyrot resistance phenotype. In addition, speculate on whether the mutations are likely to be dominant or recessive when heterozygous with the wild-type (lilyrot-sensitive) allele. (10%)

Answer 11: Taking each new mutation in turn:

Mutant 2 Nonsense mutation

   Almost certainly doesn't make a functional protein, as the receptor protein it does make will be truncated. Lilyrot-resistant. Likely to be recessive. However, if the functional receptor acted only as a dimer, then the truncated receptor protein might heterodimerize with the wild-type receptor protein and this heterodimer may be non-functional. In this case, the mutant allele might show intermediate dominance.

Mutant 3 A to G change in the promoter region

   Depends entirely on whether this affects the promoter binding site.

   If it doesn't affect the binding site then it will be wild-type.

   If it stops RNA polymerase binding then it's a null (no receptor protein is produced), lilyrot-resistant and almost certainly recessive.

   It could result in less expression, making less receptor protein, maybe slowing down infection progression rate. This would be recessive or possibly show intermediate dominance.

   It could result in more expression, increasing the number of receptors, which may strongly (dominant) or partially (intermediate dominance) or negligibly (neutral) quicken infection rate.

Mutant 4 Missense mutation

   Depending on the specific amino acid change and its position in the protein:

   No effect on protein function.

   Decreases receptor protein function or eliminates protein function, or changes binding site so virus cannot bind, or stops protein entering the membrane any of which might generate a loss-of-function lilyrot-resistant phenotype where the mutant allele is recessive.

   Increases protein function or longevity (by decreasing degradation rate), which may make a hypersensitive mutant that would be dominant.

Mutant 5 3 base pair insertion into the coding region.

   This isn't a frame-shift. So this could have any of the affects seen in mutant 4 depending on whether receptor protein function is affected or not.

Question 12: Now that you understand the genetic basis of colour, leaf shape and lilyrot resistance, how would you generate lilyrot-resistant plants (a) by creating genetically modified, transgenic lilies and (b) without creating genetically modified, transgenic lilies? (10%)

Answer 12: Taking each question in turn:

(a) Any scheme that generates transgenic lilies in which the receptor gene is mutated (such as a targeted insertion that interrupts the gene) or reduces expression (such as RNAi targeted at the receptor gene) should produce resistant lilies. Other strategies are also possible.

(b) Any crossing scheme that generates a strain that is homozygous for the wild-type savage gene (S^+/S^+) and homozygous wild-type for the mutant receptor gene (L^r/L^r), would produce resistant lilies.

Chapter 2.5: Strawberry Fields Forever

Question 1: The DNA that is transferred from the recombinant Ti plasmid to the strawberry germline has been designed to address the issue of slow growth and fruit colour. Draw a cartoon of the piece of DNA that is transferred to the host plant (coloured grey and black in Fig. 2.5.4) clearly stating the function of each section of the DNA being transferred. (10%)

Answer 1: The simplest explanation of the two problems (yellow colour and slow growth) in *Fragaria esculentus* var. *crundalli* is that the strain carries two mutations, one in the fruit pigment pathway (so the fruit are yellow rather than red) and the other in a growth hormone pathway (so the plants are shorter than normal). Therefore, within the T-DNA the construct must contain two genes: the wild-type copy of the fruit colour gene that is mutant in the host, and a wild-type copy of the growth hormone gene that is similarly mutant in the host. Of course there are mutations in other genes that could cause the same phenotypes, for example genes encoding proteins that perhaps transfer pigments, but key to the answer is that the two transgenes must be the wild-type alleles of the two genes that are mutant in *Fragaria esculentus* var. *crundalli*. Clearly the mutant alleles at the two genes in *Fragaria esculentus* var. *crundalli* will both have to be recessive, and the wild-type alleles in the two transgenes will have to be dominant.

Question 2: Name two distinctly different ways in which you could determine that the new DNA had arrived in its new strawberry host? (10%)

Answer 2: You could test by:
Phenotype if the plants have red fruit and normal growth.
Direct detection of transgene DNA – amplify a short DNA sequence that is unique to the transgene by PCR (or any other molecular genetic technique that detects DNA that is unique to the construct).

Question 3: How would you identify the DNA sequence flanking the insertion site of the modified T-DNA in its new strawberry host? (5%)

Answer 3: You could sequence the entire genome of the host organism and, as you know the sequence of the transgene ends, search for the sequence adjacent to them. Alternatively, you could undertake 'plasmid rescue', where the host genome is cut by a restriction enzyme and all resulting fragments are circularised using a ligase and are then transformed into a host such as *Escherichia coli*. If the transgene contains a resistance gene (such as Km^R) and the host is Km^S, then the only hosts to survive must have the plasmid with the transgene and this contains flanking endogenous DNA. If you sequence the plasmid, including the endogenous flanking sequence, you can find the insertion site by looking at the host's entire DNA sequence and seeing which sequence is homologous.

Question 4: Suggest two possible explanations for strain Reg-C having orange fruit. (10%). How would you distinguish between these explanations?

Answer 4: As all the other transgenic lines Reg-A, B, D, E, F, which have an insertion of the same T-DNA, are red, the orange colour of Reg-C must be a consequence of its insertion site. Two plausible explanations are:

insertion into a pigment pathway gene that reduces flux through the pathway;

insertion near local repressor element inhibiting expression of the transgene.

You could distinguish between the two possibilities by investigating the insertion site and determining whether the transgene insertion has mutated an endogenous gene that affects flower colour.

Question 5: Your first thought is that the different transgenic strains may have different transgene copy numbers. How could you investigate this theory? (5%)

Answer 5: You need to calculate the relative dosage of transgene DNA against genomic DNA. Whole genome sequencing will allow you to do this, as it enables you to calculate the number of times a specific sequence is encountered for sequencing. A single homozygous transgene insertion would be present at the same frequency as any endogenous genomic sequence. This could also be tested using qPCR on a DNA substrate. You could distinguish between these two possibilities by investigating the insertion site and determining whether the transgene insertion has mutated an endogenous gene that affects flower colour.

Question 6: Speculate on how a single insertion of the same transgene at different positions in identical host genomes might result in the different growth rate phenotypes seen in Fig. 2.5.5. (10%)

Answer 6: Transgene expression is often influenced by the chromosomal landscape at which it has been inserted. If the transgene lands in a transcriptionally active site, it may be more highly expressed than if it lands at a transcriptionally silent or even transcriptionally repressive site. The way in which local endogenous control elements affect expression of a transgene is very complex and will also depend on the specific sequences, topology and control elements within the transgene.

Question 7: Briefly describe how you would ensure that that the taste test produces reliable results? (5%)

Answer 7: In any test where the participant (in this case the customer) could be influenced by things other than those intended, it is important to eliminate all extraneous variables. Here, the strawberries would need to be tasted in the absence of other sensory cues, such as vision. Ideally therefore, in this case the customer should at least be blindfolded. These are often called blind tests.

Question 8: Speculate why there are two different classes of Reg-B strain. (10%)

Answer 8: The most plausible explanation is that the transgene has inserted into an essential endogenous gene. Therefore you cannot get individuals that are homozygous for the transgene. From a genetic perspective this behaves like a classic recessive lethal mutation. The slow plants have no transgene (and are in fact the original unmodified strain), whereas the fast plants are heterozygous for the transgene insertion. Individuals homozygous for the transgene insertion die. To take each cross in turn:

slow × slow: the original strain with no transgene, true-breeding – these will have yellow fruit;

fast × slow: fast-growing strawberries have one copy of the transgene (heterozygous), have red fruit and are crossed to the original slow-growing non-transgenic yellow fruit plants to generate an F1 with equal numbers of fast-growing 'heterozygous' transgenic red-fruiting plants and original slowing growing non-transgenic strawberries;

fast × fast: crossing two red-fruiting strawberries, both of which are fast growing with red fruit and a single insertion should give a 1:2:1 ratio of no transgenes (original slow yellow strain), one transgene (fast-growing with red fruit) and two transgenes (homozygous for the transgene and presumably having red fruit and being very fast growing) – in this case you have 1 slow (homozygous for no insertion), 2 fast (heterozygous for the insertion, and 1 dead (homozygous for the insertion). So it is a 1:2:1 ratio but you only see a 1:2 ratio.

Question 9: How might you generate a true-breeding, fast-growing Reg-B strain? (10%)

Answer 9: You could add a second transgene to the Reg-B strain that contains a wild-type copy of the gene that is interrupted in the Reg-B strain. In reality, you would initially put this novel transgene into a second strain and cross it into the Reg-B transgenic line.

Question 10: How does qPCR enable you to investigate gene expression levels? (5%)

Answer 10: qPCR amplifies from an mRNA template, so the rate at which a specific mRNA is amplified reflects the original level of that mRNA template.

Question 11: What is the relationship between *giant* expression and plant growth rate? (5%)

Answer 11: Generally, the higher the expression of *giant* in the transgenic strawberries, the faster they grow. So there would seem to be a very close correlation between *giant* expression and growth rate.

Question 12: Speculate on the mechanism by which *giant* expression is affected by the presence of the transgene introduced by Professor Marie Nation. How would you test this? (15%)

Answer 12: The gene used by Professor Nation in the transgene must lie functionally upstream of the *giant* gene as the presence of the transgene seems to change *giant* gene expression. This could be direct (perhaps the transgene contains a transcriptional factor that activates *giant* expression), or has a similar consequence by a less direct route (perhaps the transgene contains a protein that degrades a repressor of giant expression). This could be tested by looking at direct interactions, making mutations in the *giant* gene, or a variety of other routes.

Chapter 2.6: The Mystery of Trypton Fell

The questions in this chapter first appeared in a Level 3 Genetics exam at the University of Glasgow in December 2013.

Question 1: What is the evidence that the pigs are dying from an X-linked genetic condition? (10%)

Answer 1: Looking at Fig. 2.6.1, it's clear that only males are affected. In addition, Porky (I:2) passes the condition to his grandsons (III:6 and III:9) via his daughter (II:6), who must therefore be a carrier. There are no father-to-son transmissions. An unaffected father within the core family never passes the condition onto his offspring nor any subsequent generation. Clearly this is a very limited dataset and you must therefore treat any interpretation with caution. Nevertheless, X-linkage is a plausible explanation.

Question 2: Lady Mary tells you that female II:6 suffered from occasional unexplained paralysis, but died at a respectable old age after an unfortunate incident with a runaway tractor. Interestingly one of her daughters (III:7) also exhibited occasional unexplained paralysis. Why might female II:6 and her daughter III:7 have suffered from occasional unexplained paralysis? (10%)

Answer 2: If it is an X-linked condition then because of X-inactivation, carrier females will be a mixture of wild-type and mutant cells. Therefore you would expect only about half of the cells in a carrier female to be wild-type at this gene (as their mutant X-chromosome is condensed into a Barr body), whilst the other half uses the mutant version of the gene (as their wild-type X-chromosome is condensed into Barr body). Depending on the way in which the gene is normally expressed, carrier females may have low expression of this gene and therefore have a mild phenotype. In this example, it is possible that 'temporary paralysis' is a mild version of sudden unexplained death. A paralytic phenotype suggests that you are probably looking at a mutation that affects muscle or neural function.

Question 3: Speculate on the genetic basis of coat colour in Lady Mary's pigs. (10%)

Answer 3: Male pigs exist in two colour forms (black or pink), whereas females can be black, pink or 'blotchy', which is a random mixture of black and pink. The most plausible explanation is that coat colour is controlled by a gene on the X-chromosome. Let's suppose there are two alleles, pink and black. Males have one X-chromosome and have either the pink allele (so are pink) or have the black allele (so are black). Females have two X-chromosomes and therefore may have two pink alleles (so are pink), or two black alleles (so are black) or have one of each (a pink allele and a black allele). In the latter case, the females appear 'blotchy' as a consequence of X-inactivation so some patches of skin use the pink allele and some patches use the black allele.

Question 4: Using appropriate genetic notation, draw Professor Trotter's crossing scheme, showing the genotype of each individual. (10%)

Answer 4: Let's define the genetic notation as:

death gene has two alleles resistant (sus^+) is dominant to susceptible (sus^-):
ear gene has two alleles normal (rag^+) is dominant to ragged (rag^-);
colour gene has two alleles black (col^B) and pink (col^P).
For the colour gene, heterozygous females are blotchy, so you shouldn't use dominant and recessive as descriptors.

Note that any appropriate genetic notation is fine.

You should draw something along the following lines:

$$\frac{sus^- \ col^B \ rag^+}{Y} \times \frac{sus^+ \ col^P \ rag^-}{sus^+ \ col^P \ rag^-}$$

(male 'Porky' *susceptible* × female 'Peppa' *pink* and *ragged*)

$$\frac{sus^- \ col^B \ rag^+}{sus^+ \ col^P \ rag^-}$$

F1 wild-type female as above (and heterozygous at each gene)

Second cross

$$\frac{sus^- \ col^B \ rag^+}{sus^+ \ col^P \ rag^-} \times \frac{sus^? \ col^? \ rag^?}{Y}$$

In this cross the F1 female is triply heterozygous. The male genotype doesn't matter (as you are only scoring the genotypes of his sons; he will be simply donating the Y-chromosome to his sons, so the genotype of his X-chromosome is irrelevant), which is why they are drawn as $sus^?$ $col^?$ $rag^?$, meaning the *sus*, *col* and *rag* alleles are unknown.

The F2 male offspring all have the Y-chromosome from their father and an X-chromosome from their mother. There are eight possible X-chromosomes (and eight possible phenotypes) and these are shown here.

sus^+	col^B	rag^+
sus^-	col^P	rag^-
sus^+	col^B	rag^-
sus^-	col^P	rag^+
sus^+	col^P	rag^+
sus^-	col^B	rag^-
sus^-	col^B	rag^+
sus^+	col^P	rag^-

Note that it's important to draw the crossing scheme in such a way that it's clear which alleles are on which X-chromosome.

Question 5: It is stated that the F1 triply heterozygous females can be mated to any random males. Why doesn't the genotype of the males matter and why do you only use their F2 sons to calculate the map distance? (10%)

Answer 5: You only use the F2 males to calculate the map distance as their X-chromosome is derived from their triply heterozygous mother, so their phenotype absolutely reflects their genotype. As their father only contributes a Y-chromosome, it doesn't matter what X-chromosome genotype the father has.

Question 6: Calculate the map distances between the three X-linked genes, and draw a map to show their relative positions. (10%)

Answer 6: The commonest classes of offspring are the parental types (susceptible black normal, and resistant pink ragged), whilst the rarest types are the double cross-overs (susceptible pink ragged, and resistant black normal), which tells us that the *susceptible* (*sus*) gene is in the middle of the three. The map distances are:

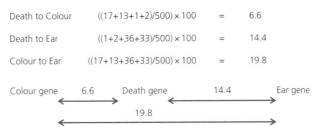

Death to Colour	$((17+13+1+2)/500) \times 100$	=	6.6
Death to Ear	$((1+2+36+33)/500) \times 100$	=	14.4
Colour to Ear	$((17+13+36+33)/500) \times 100$	=	19.8

Colour gene ←— 6.6 —→ Death gene ←— 14.4 —→ Ear gene
←—————————— 19.8 ——————————→

Question 7: How could you find the DNA sequence of the mutant sudden unexplained death gene within the pig X-chromosome? What problems might you encounter? (10%)

Answer 7: If, as implied, the region within which the sudden death gene lies is fairly imprecise and therefore large and containing many genes, it may be sensible to narrow the region of interest by further gene mapping. Sequencing and looking for obvious mutations (nonsense or frameshift) might indicate which genes are worth focusing on. Similarly, looking for candidate genes, genes that have previously been described that could plausibly cause a paralytic phenotype when mutant, may help. However, there are likely to be many missense mutations in putative coding regions and changes in potential control regions, the consequences of which are very difficult to determine precisely. Without significant further experimentation it may be difficult to decide which DNA change causes the mutant paralytic effect.

Question 8: How are you going to prove which, if any, of the sodium channel genes, is the cause of the mutant phenotype? What problems might you encounter? (15%)

Answer 8: You could look at the sequences of each of the three sodium channel genes and see whether susceptible mutant pigs have an obvious mutation within any of them. This could be a nonsense mutation or frame-shift mutation, or a deletion or insertion that affects the coding region. Other DNA changes, such as missense mutations or non-coding changes adjacent to the coding region could cause a phenotype but they may be asymptomatic or cause a severe phenotype, or anything in-between, so whether they are the cause of this mutant phenotype or not is more difficult to determine. Clearly, you are looking for a 100% correlation between a mutation in the DNA and the mutant phenotype. You might be able to prove which gene is responsible for the mutant phenotype by gene rescue. Here the key is to get a wild-type transgene into a mutant pig and seeing whether it restores, or at least partially restores, the mutant phenotype. The best approach would be to put the transgene into a wild-type embryo and cross it into the mutant background (non-homologous recombination approach, so the transgenic pig carries an extra copy of the gene) or put the wild-type transgene into a mutant embryo (homologous recombination resulting in gene replacement). As there are three candidate genes, this would need to be repeated with each gene individually. Problems with the non-homologous approach would be getting the spatial and temporal expression of the gene correct, as the control region may be big and/or complex and therefore difficult to include in a transgene. So, one of the key points here is that it's not easy. A similar approach could be made using genome-editing technology such as CRISPR.

Question 9: Who are the parents of (a) Lady Mary, (b) Lady Edith, (c) Lady Sybil, (d) Keanu Bottomless-Pitt, (e) Chardonnay Bottomless-Pitt. (5%)

Answer 9: In every case, for each and every STR the child must receive 1 STR allele from their mother and 1 STR allele from their father. Therefore, overall they will receive 13 STRs from each parent. It follows that the parents are:: (a), (b) & (c) Lord and Lady Trypton; (d) Lord Trypton and Sharon Bottomless-Pitt; (e) Darren and Sharon Bottomless-Pitt.

Question 10: Darren and Sharon Bottomless-Pitt are not actually married. Explain why they may share the same family name? (5%)

Answer 10: Excluding STR13 (which seems to be co-inherited with the sex of an individual where the 14 allele appears to be on the X-chromosome and the 19 allele appears to be on the Y-chromosome), Darren and Sharon share 6 STRs so are probably cousins, meaning their fathers are probably brothers.

Question 11: Why might Lord Trypton be concerned at the relationship between Lady Sybil and Keanu Bottomless-Pitt? (5%)

Answer 11: They are half-siblings, because Lord Trypton is the father of both of them.

Chapter 2.7: Sir Henry's Enormous Chest

Question 1: What is the evidence that NPS is an autosomal dominant condition? (5%)

Answer 1: As NPS is in every generation of the family and only occurs in someone who has an affected parent, it is very likely to be dominant. If NPS was recessive, the presence of affected individuals in generations II and III would only occur if they inherited mutant alleles from both parents. This would mean that genetically unrelated individuals I.1 and II.1 would both have to be heterozygous, and I.2 would be homozygous which is most unlikely in a rare condition. Again this suggests NPS is dominant. If NPS was sex-linked recessive, all the sons in generation II would inherit the disease from their homozygous mother, but they do not. If it was a sex-linked dominant, females III.3, III.4 and III.5 would get the disease gene on the X-chromosome inherited from their father. They do not have the condition, so it cannot be sex-linked dominant either.

Question 2: Redraw Fig. 2.7.3 showing, wherever possible, the ABO genotypes for all members of the pedigree. (10%)

Answer 2: Figure 2.7.3 is reproduced here to show the ABO blood group genotypes. For clarity blood groups shown in red are from affected individuals.

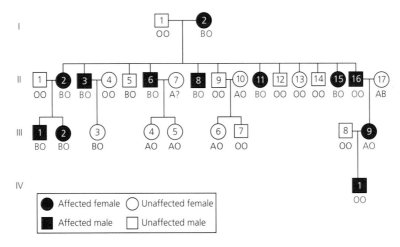

Question 3: Are there any individuals whose ABO genotype cannot be established? If so, who are they, why can their ABO status not be determined, and how could you resolve this? (10%)

Answer 3: Individual II:7 may be AA or AO. She has the 'A' blood group so must have an A allele and she passes an A allele to each of her 'A' blood group children who inherit an O allele from their father (the children must be AO). This pattern of inheritance is equally valid irrespective of whether individual II:7 is AA or AO, because both are compatible with being blood group 'A' and passing on an A allele to two children. You could resolve this by sequencing both of II:7's ABO genes to establish whether they are the same (AA) or different (AO). To be absolutely confident you should also acquire the DNA sequence of the ABO genes of her close relatives and follow them by descent.

Question 4: What is the evidence that NPS is genetically linked to ABO? (5%)

Answer 4: Blood group B and NPS seem to be inherited together. Looking at generation 1, the dominant NPS mutation is carried by female I:2 who has a BO genotype. Of her 12 children, 6 have inherited the B allele (5 of whom are affected by NPS and 1 who is not) and 6 have inherited the O allele (1 of whom is affected and 5 who are not). This strongly suggests the dominant NPS mutant allele is closely linked to the B blood group allele.

Question 5: Calculate the map distance between NPS and ABO. (10%)

Answer 5: The offspring of parents that are heterozygous for both ABO and NPS can be used to calculate the map distance. There are 18 of these, 12 in generation II, 5 in generation III and 1 in generation IV. In the figure the status of each individual is marked in blue as parental (P) or recombinant (R). For clarity, blood groups shown in red are from affected individuals.

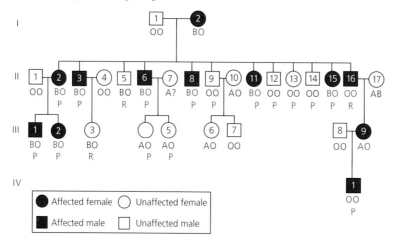

There are three recombinants from 18 offspring: 3/18 = 16.7%.

Note that the parents II:9 and II:10 are both unaffected by NPS (so they are both homozygous for the wild-type NPS allele); therefore, you cannot judge whether recombination occurs or not using phenotype data.

Similarly, affected father II:16 has the 'O' blood group so is homozygous OO and you cannot tell whether recombination has occurred or not.

Question 6: If male III:1 from the pedigree and an unaffected woman with blood group O have a child, what is the probability that it will be a boy, have blood group B and have nail–patella syndrome? (5%)

Answer 6: Half (0.5 or 50%) will be male, half (0.5 or 50%) will be blood group B and, as NPS is linked to the B allele in this male with a map distance (recombination frequency) of 16.67, then

83.3% (or 0.833) will co-inherit the NPS allele. Therefore the probability of a male, B, NPS child is $0.5 \times 0.5 \times 0.833 = 0.208$. Or if you prefer this in percentage terms, $50\% \times 50\% \times 83.3\% = 20.8\%$, though it's much harder to calculate this in percentages.

Question 7: If you found another pedigree apparently showing inheritance of a condition with an NPS-like phenotype, would you be justified in combining the data from the two families? What might make you hesitate over such a decision? (5%)

Answer 7: Note the doubt suggested by 'apparently' in the first line of the question. The mutation in the new pedigree might be in a different gene giving the same phenotype.

Question 8: Speculate on how a single base-pair change in the coding region of the NPS transcription factor gene could result in a dominant mutation leading to the NPS mutant phenotypes. (5%)

Answer 8: There are several possible answers. The most plausible explanation is a 'gain of function' (GOF) mutation. One possibility is that a missense mutation changes one amino acid in the transcription factor (TF) that results in it being more active or more stable. This enhanced TF function would override the effect of the other wild-type allele. If the NPS TF acts to increase gene expression, a missense mutation may mean that it is 'irrepressible', binding to a control region of a gene but being unable to be removed (this is similar in principle to some mutations in the *lac* operon). Alternatively, the NPS TF may be a transcriptional repressor, where a simple loss-of-function mutation means it cannot repress its target gene, which as a result is over-expressed.

Question 9: Speculate on how a single base-pair change in a non-coding region of the NPS transcription factor gene could result in a dominant mutation leading to the NPS mutant phenotypes. (5%)

Answer 9: There are several possible answers. As the mutation doesn't affect the coding region it can really only affect control elements thus altering aspects of gene expression (probably transcription) or splicing. Focusing on changes in gene expression, single base-pair changes in control (promoter or enhancer) elements could affect the binding of transcription factors thus altering gene expression. Here as the mutation is dominant the most plausible (though by no means the only) answer is inappropriate expression, meaning the gene is expressed at a significantly higher level than normal or is expressed at a time or place (spatial or temporal expression) where it is not normally expressed. For example this could be the loss of a transcription repressor binding site, so the target gene cannot be repressed. Any of these could lead to a dominant phenotype.

Question 10: Speculate on why the same mutation in different members of the same family causes a different severity of phenotype in different individuals. (10%)

Answer 10: No gene functions in glorious isolation, and few phenotypes are entirely associated with variation at a single gene. It's very likely that variation at other genes affects the severity of phenotype associated with NPS. The second site mutations are usually called genetic modifiers. They can increase or decrease the effect of the original mutation, usually by some functional interaction. So if there is variation at so-called modifier genes this may cause great variety in phenotype even in those individuals that have the same causal mutation. More specifically, here you know that NPS encodes a transcription factor (TF), so variation at other TFs that interact with NPS TF to increase or decrease expression of target genes may affect what happens when the NPS TF gene is mutant.

Question 11: Why might standard diagnostics tests only identify 85% of affected families? (5%)

Answer 11: Within a population, some NPS mutations will be more common than others, so standard diagnostic tests are probably designed to identify these common mutations. New or rare mutations within the coding sequence are easy to find if they have a severe effect on the protein (nonsense mutations that generate truncated proteins, or insertions or deletions that cause frame-shift mutations). However, missense mutations (where one amino acid is substituted for another) may or may not affect protein function, depending on the biochemical properties of the two amino acids and their position within the protein. In addition, mutations outside the coding region could affect gene expression but these are also difficult to prove as causal.

Alternatively, there could be another gene that, when mutant, causes a condition that is more-or-less indistinguishable from NPS.

Question 12: Speculate on why wild-type mice carrying the human NPS mutant transgene do not have a mutant phenotype. (5%)

Answer 12: The mouse and human NPS gene homologues may have diverged significantly over evolutionary time resulting in divergence of gene function. Therefore, the human gene may have no function or consequence in the mouse. In this specific example, the NPS gene encodes a transcription factor, so the specific DNA sequence it binds to in humans may be different in mice. Without further investigation it is difficult to say which differences between human genes and mouse genes actually have a consequence for gene function.

Question 13: Prior to introducing it to a wild-type mouse embryo, what modifications should they make to the mouse NPS gene that would enable them to investigate why the original NPS mutation identified by Renwick is dominant. What problems might they encounter? (10%)

Answer 13: If the mouse gene functions in a broadly similar way to the human gene, it may be better to modify the mouse NPS gene by introducing the human mutation. The problem here is that the mutation, at least in humans, is dominant. Therefore transgenic mice carrying the mutant gene will have a mutant phenotype, even though they may only carry one transgenic copy of the gene, and this could affect the viability or fitness of the host mouse. Note that this is not normally a problem when creating models of recessive mutations, as you have to mutate both homologues to have a mutant phenotype, so the transgenic recessive mutation can be kept as an unaffected heterozygous stock.

Question 14: Given that you now have a transgenic mouse model, what experimental strategy would you use to find a treatment for human NPS? (10%)

Answer 14: There are lots of possibilities. As the mutant phenotype is caused by a dominant mutation transcription factor, you might want to investigate which gene(s) the wild-type and mutant versions the mouse NPS transcription factor binds upstream of to affect transcription levels. You might use DNA binding assays and/or microarrays comparing wild-type vs NPS in the affected tissues. As it is a rare condition, it is reasonable to assume that most people affected by NPS are heterozygous, so a successful treatment strategy might involve somehow destroying the mutant NPS transcription factor or transcript, without affecting the wild-type NPS transcription factor or mRNA. The latter might be achieved using an RNAi approach that flags the dominant mutant mRNA for degradation but leaves the wild-type unaffected. The two key issues here would be delivering the dsRNA to the correct cell type(s) and ensuring the specificity of the dsRNA (degrading the mutant mRNA but not the wild-type mRNA). In humans this would be very difficult to achieve.

Chapter 2.8: Pandemonium

Question 1: Who is the father of the infant? What is the evidence for this? (10%)

Answer 1: Niall is the father. At each STR the infant receives one repeat allele from its mother and one from its father, in a standard Mendelian fashion. As we know the STR profile of the mother (in yellow) and the infant (in orange), the father must provide all the STR alleles that the mother does not. As Niall (red) provides the entire paternal profile of the infant (orange) he must be the father. The other males only match none, one or two STR alleles, (green) which is a standard background population level of similarity.

STR	Jane	Harry	Liam	Louis	Niall	Zayn	Infant
STR A	7/10	8/9	12/15	11/13	9/10	7/11	7/10
STR B	11/13	12/17	14/15	12/15	12/16	10/13	11/16
STR C	26/31	24/24	29/30	28/31	25/27	26/30	25/31
STR D	21/23	19/23	24/26	20/25	22/25	25/26	22/23
STR E	24/25	28/31	24/29	27/29	25/30	26/29	25/30
STR F	16/17	12/17	14/14	15/18	13/14	12/18	14/16
STR G	23/23	19/22	20/24	21/22	21/23	19/24	21/23
STR H	8/14	11/11	9/12	10/13	12/13	8/9	12/14
STR I	32/33	31/38	34/35	36/36	34/37	35/39	33/37
STR J	8/12	8/9	10/13	9/10	11/14	12/15	11/12

Question 2: Why might giant pandas from a different Chinese province be a valuable addition to the Chengdu breeding programme? (10%)

Answer 2: Outbreeding good, inbreeding bad.

Question 3: Taking each female in turn (and at this stage ignoring the males), discuss which female(s) may be Yin Yang's mother? (5%)

Answer 3: The best way to approach this is to determine who cannot be the mother of Yin Yang by simply finding STR alleles in a female that are not found in Yin Yang.

Female 1 is not the mother due to STR 5 (she is 7/9 whereas Yin Yang is 8/8), and STR 10 (she is 17/17 whereas Yin Yang is 19/20)

Female 2 could be the mother.

Female 3 is not the mother due to STR 5 (she is 7/7 and Yin Yang is 8/8), and STR 13 (she is 14/14 whereas Yin Yang is 12/13).

This is summarised in the table here; STR alleles that are not passed to Yin Yang are shown in red.

Name	Female 1	Female 2	Female 3	Male 1	Male 2	Male 3	Yin Yang
STR 1	6/7	6/6	6/8	7/7	7/8	6/6	6/7
STR 2	11/11	12/13	12/13	11/12	12/12	11/13	11/13
STR 3	22/23	23/25	23/24	23/23	23/24	22/24	23/25
STR 4	10/11	9/10	11/11	11/12	10/11	9/11	9/11
STR 5	7/9	8/9	7/7	7/8	8/9	8/9	8/8
STR 6	15/17	15/17	14/16	16/16	15/15	14/15	15/16
STR 7	19/21	22/22	21/22	19/21	21/21	20/21	21/22
STR 8	15/18	15/16	15/15	15/18	17/18	15/17	15/15
STR 9	9/10	8/10	9/10	8/10	10/10	9/10	10/10
STR 10	17/17	19/20	20/20	17/19	18/20	18/20	19/20
STR 11	5/5	5/6	5/7	6/6	6/6	6/7	5/6
STR 12	17/17	17/17	17/17	17/22	17/22	17/22	17/22
STR 13	12/12	12/12	14/14	11/12	11/14	13/14	12/13

Question 4: Taking each male in turn (and at this stage ignoring the females), discuss which male(s) could be Yin Yang's father? (5%)

Answer 4: Again, the best way to approach this is to determine who cannot be the mother of Yin Yang by finding STR alleles in the potential fathers that are not found in Yin Yang.
Male 1 could be the father.
Male 2 is not the father due to STR 2 (he is 12/12 whereas Yin Yang is 11/13), STR 8 (he is 17/18 whereas Yin Yang is 15/15), and STR 13 (he is 11/14 whereas Yin Yang is 12/13)
Male 3 is unlikely to be the father due to STR 3 (he is 22/24, whereas Yin Yang is 23/25)
Therefore the father of Yin Yang could be male 1.

This is summarised in the earlier table; STR alleles that are not passed to Yin Yang are shown in red.

Question 5: Which pair, if any, might be Yin Yang's parents? What issues, if any, do the results from the STR profiling raise? (10%)

Answer 5: From the argument put forward earlier, it's tempting to think that female 2 is the mother and male 1 is the father, but STR 13 makes this unlikely as the potential mother is STR 13 (12/12), the father is STR 13 (11/12) and Yin Yang is STR 13 (12/13). Therefore, this would require a mutation. Mutations are possible, as STRs mutate at a higher frequency than unique sequences. An alternative possibility is that other as yet unidentified giant pandas are living in Shaanxi and one or two of these are the are the parents of Yin Yang. These may, or may not, be close relatives of female 2 and male 1.

Question 6: Speculate on whether any of the STRs could be used as a sex test? What sex does this suggest Yin Yang is? (10%)

Answer 6: Giant pandas have the standard mammalian XX female XY male sex-determining system. Therefore, you are looking for an STR that is found on the sex-chromosomes and is always a specific size on an X-chromosome, and always a specific (but different) size on a Y-chromosome. The STR that conforms to that pattern is STR 12. All the known females (XX) have an STR 12 genotype of 17/17 whereas all known males (XY) have an STR 12 genotype of 17/22. It follows that STR 12 allele 17 could be on the X-chromosome, whereas STR 12 allele 22 could be on the Y-chromosome. By these criteria Yin Yang, with STR 12 genotype 17/22, is male (XY). He has the STR 12 allele 17 from his X-chromosome and STR 12 allele 22 from his Y-chromosome.

Note that in human DNA profiling the amelogenin gene is used to determine sex. The amelogenin gene is found on both the X-chromosome and the Y-chromosome, but the X-chromosome version has a 6 bp deletion in intron 1 relative to the Y-chromosome version. Therefore, the X-chromosome amelogenin PCR product is 6 bp shorter than the Y-chromosome amelogenin PCR product.

Question 7: Who is the father of the new cub? What is the evidence for this? (10%)

Answer 7: Taking each male in turn:

Male 1 is not the father due to STR 3 (he is 23/23 and although the new cub is 23/24 it must have received the 23 from its mother as she is 22/23, so the cub must receive a 24 from its father), STR 6 (he is 16/16 whereas the cub is 15/17), and STR 10 (he is 17/19 and although the cub is 17/20 it must have received the 17 from its mother as she is 17/17, so the cub must receive a 20 from its father), and STR 13 (he is 11/12 and although the cub is 12/14 it must have received the 12 from its mother as she is 12/12 so the cub must receive 14 from its father).

Male 2 is not the father due to STR 9 (he is 10/10 and although the new cub is 8/10 it must have received the 10 from its mother as she is 9/10 so the cub must receive an 8 from its father).

Male 3 is not the father due to STR 2 (he is 11/13 and although the new cub is 11/12 it must have received the 11 from its mother as she is 11/11, so the cub must receive 12 from its father), STR 8 (he is 15/17 and the cub is 18/18) and STR 9 (he is 9/10 and although the cub is 8/10 it must have received the 10 from its mother as she is 9/10 so the cub must receive a 8 from its father).

Male 4 is not the father due to STR 9 (he is 9/10 and although the cub is 8/10 it must have received the 10 from its mother as she is 9/10 so the cub must receive a 8 from its father) and STR 13 (he is 11/12 and although the cub is 12/14 it must have received the 12 from its mother as she is 12/12 so the cub must receive 14 from its father).

Male 5 could be the father.

This is illustrated in the table below where STR alleles that cannot be passed to the cub from a potential father are shown in red.

Sex	Female 1	Male 1	Male 2	Male 3	Male 4	Male 5	Cub
STR 1	6/7	7/7	7/8	6/6	7/7	7/8	6/7
STR 2	11/11	11/12	12/12	11/13	11/12	12/13	11/12
STR 3	22/23	22/23	23/24	22/24	22/24	23/24	23/24
STR 4	10/11	11/12	10/11	9/11	10/10	11/11	10/11
STR 5	7/9	7/8	8/9	8/9	7/8	8/8	8/9
STR 6	15/17	16/16	15/15	14/15	15/17	15/16	15/17
STR 7	19/21	19/21	21/21	20/21	19/20	20/21	19/21
STR 8	15/18	15/18	17/18	15/17	17/18	16/18	18/18
STR 9	9/10	8/10	10/10	9/10	9/10	8/9	8/10
STR 10	17/17	17/19	18/20	18/20	18/20	19/20	17/20
STR 11	5/5	6/6	6/6	6/7	6/7	5/6	5/6
STR 12	17/17	17/22	17/22	17/22	17/22	17/22	17/22
STR 13	12/12	11/12	11/14	13/14	11/12	13/14	12/14

Question 8: What sex is the new cub? What is the evidence for this? (5%)

Answer 8: The cub is male as it the STR 12 genotype 17/22 which you have already determined is characteristic of a male (see answer to question 6).

Question 9: Why is it more difficult to determine parentage in Giant Pandas than in Chimpanzees? (10%)

Answer 9: Simply because giant pandas are very inbred, which means there isn't a lot of variation in their STRs nor any other part of their genome.

Question 10: What experiment could you do to determine the chromosomal number of a cell and, in particular, how could you identify which chromosome is present in abnormal number? (10%)

Answer 10: A simple chromosome squash would be a feasible approach; there are many ways in which different chromosomes can be labelled with different stains. Looking for Barr bodies would indicate the number of X-chromosomes present in a cell. Next generation sequencing can also be used to determine relative dosage of DNA, as can qPCR on a DNA substrate.

Question 11: What name is given to the human genetic XXY condition? What phenotype do human XXY males have? (5%)

Answer 11: Kleinfelter's syndrome. Those affected are usually sterile and may have other medical issues, which may be quite variable and are often quite mild.

Question 12: Speculate on the cause of the XXY karyotype of Yin Yang. In particular can you determine whether non-disjunction occurred at meiosis I or at meiosis II of his mother or his father? (10%)

Answer 12: Yin Yang must have inherited two X-chromosomes, both carrying the 42 repeat STR, from his mother and received a Y-chromosome from his father. Non-disjunction must have occurred during meiosis II, in a cell carrying the 42 repeat-bearing X-chromosome consisting of two chromatids, both of which went into the egg.

Acknowledgement

I'd like to thank Alaina Macri from Edinburgh Zoo for advice during the preparation of this chapter.

Further information

Information on the Chengdu Research Base of Giant Panda Breeding website can be found at http://www.panda.org.cn/english/ (accessed July 2015).

Further reading

Li, R., Fan, W., Tian, G. *et al.* (2010) The sequence and *de novo* assembly of the giant panda genome. *Nature* **463:** 311–17.

Zhao, S., Zheng, P., Dong, S. *et al.* (2013). Whole genome sequencing of giant pandas provides insights into demographic history and local adaptation. *Nature Genetics* **45:**67–71.

Chapter 2.9: My Imperfect Cousin

The questions in this chapter give you an opportunity to investigate slightly more complex crossing schemes and ratios, as well as using chi-squared tests to confirm or reject your hypotheses.

Question 1: Propose a theory that explains why the offspring of a cross between strain 1 (red round) and strain 2 (yellow oval) produces F1 tomatoes that are all red and round. (5%)

Answer 1: The simplest answer is that there are two genes, one conferring colour (red vs yellow) and the other conferring shape (round vs oval). As the F1 are all red and round, then at the colour gene the red allele must be dominant to the yellow allele and at the shape gene the round allele must be dominant to the oval allele.

Question 2: Propose a theory that explains the genetic basis of tomato colour and shape as suggested by data from the second cross that is shown in Table 2.9.2. (5%)

Answer 2: These data support the theory proposed earlier in the answer to question 1 that there are two genes, one conferring colour (red vs yellow) and the other conferring shape (round vs oval). As the F1 are all red and round, then at the colour gene the red allele must be dominant to the yellow allele and at the shape gene the round allele must be dominant to the oval allele. If this theory is correct you'd expect to see a 9:3:3:1 in the F2, which is precisely how the data looks. This can be drawn as a Punnett square.

At the colour gene there are two alleles, the dominant red denoted C^+ and the recessive yellow denoted C^-. At the shape gene there are also two alleles, the dominant round denoted S^+ and the recessive oval denoted S^-. The F1 parents must both have the genotypes $C^+C^- S^+S^-$, which means they can each produce 4 classes of gamete in equal frequencies, C^+S^+, C^+S^-, C^-S^+ and C^-S^-.

		F1 gametes			
		C^+S^+	C^+S^-	C^-S^+	C^-S^-
F1 gametes	C^+S^+	$C^+C^+ S^+S^+$ red round	$C^+C^+ S^+S^-$ red round	$C^+C^- S^+S^+$ red round	$C^+C^- S^+S^-$ red round
	C^+S^-	$C^+C^+ S^+S^-$ red round	$C^+C^+ S^-S^-$ red oval	$C^+C^- S^+S^-$ red round	$C^+C^- S^-S^-$ red oval
	C^-S^+	$C^+C^- S^+S^+$ red round	$C^+C^- S^+S^-$ red round	$C^-C^- S^+S^+$ yellow round	$C^-C^- S^+S^-$ yellow round
	C^-S^-	$C^+C^- S^+S^-$ red round	$C^+C^- S^-S^-$ red oval	$C^-C^- S^+S^-$ yellow round	$C^-C^- S^-S^-$ yellow oval

This resolves as a 9:3:3:1 ratio:

9 red round
3 yellow round
3 red oval
1 yellow oval

Question 3: Use Table 2.9.3 to calculate the chi-squared value to test whether your theory is plausible. What do you conclude? (15%)

Answer 3: The chi-squared value can be calculated as follows:

Phenotype	Observed number	Expected number	(O – E)	(O – E)²	$\dfrac{(O-E)^2}{E}$
red round	571	562.5	8.5	72.25	0.128
red oval	176	187.5	−11.5	132.25	0.705
yellow round	193	187.5	5.5	30.25	0.161
yellow oval	60	62.5	−2.5	6.25	0.100
total	1000	1000			1.094

As chi-squared = 1.094, and this is clearly less than 7.185 (the 5% cut-off for significance), you can accept your hypothesis that these data fit a 9:3:3:1 ratio.

Question 4: If strains 2 and 3 have mutations in the same shape and colour genes, what would be the phenotypes of their F1 and F2 offspring? (5%)

Answer 4: You have already established that strain 1 has mutations in the colour (recessive yellow) gene and shape (recessive oval) gene. Therefore if strain 2 (yellow oval: $C^-C^-\ S^-S^-$) and strain 3 (yellow oval: $C^-C^-\ S^-S^-$) have mutations in the same two genes then all their offspring (the F1) will also be yellow oval ($C^-C^-\ S^-S^-$) as will the F2 (yellow oval: $C^-C^-\ S^-S^-$). This is a complementation test.

Question 5: Use Table 2.9.5 to help you calculate the chi-squared value and test whether the data in Table 2.9.4 is in a 1:1 ratio. In this case there is one degree of freedom and the 5% cut-off is 3.84. (10%)

Answer 5: There is one degree of freedom as the only variable is the shape (round vs oval).

Phenotype	Observed number	Expected number	(O – E)	(O – E)²	$\dfrac{(O-E)^2}{E}$
yellow round	565	500	65	4225	8.45
yellow oval	435	500	−65	4225	8.45
total	1000	1000			16.90

As the chi-squared value of 16.90 is much higher than the 5% cut-off of 3.85, you reject the hypothesis that the two classes of tomato are in a 1:1 ratio.

Question 6: Speculate on any plausible non-genetic reasons why the data may deviate from a 1:1 ratio. (10%)

Answer 6: Here you are looking for effects of the environment on the two phenotypes (yellow round vs yellow oval). Clearly, from the information given so far in the story it's not possible to definitively give an answer, but the question is asking you to speculate on why, so any plausible answer will do. As there are about 25% fewer yellow oval tomatoes than yellow round tomatoes, it may be that oval tomatoes are less viable than their round siblings. As it seems unlikely that the oval shape would affect viability directly, perhaps the shape gene encodes a protein that affects a second characteristic that has a consequence on viability. In addition, if there was a viability issue, you should have picked this up in the earlier crosses. A good answer will come to the conclusion that non-genetic reasons are unlikely in this case.

Question 7: Use Table 2.9.6 to help you calculate the chi-squared value and test whether the data in Table 2.9.4 is in a 9:7 ratio. Again there is one degree of freedom and the 5% cut-off is 3.84. (10%)

Answer 7: Again there is one degree of freedom as the only variable is the shape (round vs oval).

Phenotype	Observed number	Expected number	(O − E)	(O − E)^2	$\dfrac{(O-E)^2}{E}$
yellow round	565	562.5	2.5	6.25	0.011
yellow oval	435	437.5	−2.5	6.25	0.014
total	1000	1000			0.025

As the chi-squared value of 0.025 is much lower than the 5% cut-off of 3.85, you accept the hypothesis that the two classes of tomato are in a 9:7 ratio.

Question 8: Using appropriate genetic symbols, explain why the ratio of yellow round tomatoes to yellow oval tomatoes in the F2 is 9:7. (15%)

Answer 8: Any ratio that adds up to 16 (such as the 9:3:3:1 ratio that you met earlier) suggests there are two genes involved in the phenotype(s). As all the tomatoes in this cross are yellow, you can ignore the colour genotypes as every tomato in the crossing scheme is yellow. Therefore, the most plausible explanation must be that there are two genes associated with tomato shape. Previously you said the shape gene existed in two forms: S^+ for the dominant round allele and S^- for the recessive oval allele, revealed in strain 2. From the data it seems that strain 3 behaves in the same way as strain 2, so let's suppose strain 3 has a second shape gene where T^+ is the dominant round allele and T^- is the recessive oval allele at this second shape gene. If you cross the oval parent strain 2 ($S^-S^-\ T^+T^+$) to oval parent strain 3 ($S^+S^+\ T^-T^-$) you generate an F1 population where all plants are round ($S^+S^-\ T^+T^-$). Crossing F1 ($S^+S^-\ T^+T^-$) plants to each other generates the following Punnett square.

		F1 gametes			
		S^+T^+	S^+T^-	S^-T^+	S^-T^-
F1 gametes	S^+T^+	S^+S^+ T^+T^+ round	S^+S^+ T^+T^- round	S^+S^- T^+T^+ round	S^+S^- T^+T^- round
	S^+T^-	S^+S^+ T^+T^- round	S^+S^+ T^-T^- oval	S^+S^- T^+T^- round	S^+S^- T^-T^- oval
	S^-T^+	S^+S^- T^+T^+ round	S^+S^- T^+T^- round	S^-S^- T^+T^+ oval	S^-S^- T^+T^- oval
	S^-T^-	S^+S^- T^+T^- round	S^+S^- T^-T^- oval	S^-S^- T^+T^- oval	S^-S^- T^-T^- oval

Therefore the wild-type round phenotype requires both genes to have a dominant round allele (S^+ and T^+). If either gene is mutant (homozygous recessive, S^-S^- or T^-T^-), then the phenotype is oval.

Question 9: Propose a model to explain the genetic basis of tomato shape. (15%)

Answer 9: The key point is that both shape genes need to have a wild-type allele to get a wild-type phenotype. Therefore the proteins they encode must work together, directly or indirectly, to enable the wild-type round shape. There are lots of plausible ways in which this could happen and as long as your model fully explains the data, then your model is correct. Three possible answers are:

(i) The functional protein that affects shape is a heterodimer requiring two proteins from two genes. If either gene is mutant, only one half of the protein is made and the protein is therefore non-functional.

(ii) The functional proteins are part of the same functional pathway. For example, they could encode two enzymes catalysing two steps in a biosynthetic pathway. It is only when both genes, and therefore both enzymes, function normally that the wild-type molecule at the end of the pathway can be made.

(iii) One gene encodes a transcription factor that activates the expression of the second gene that encodes the 'biologically active' protein. Again, both genes would need to be wild-type to get a functional protein.

However, alternative plausible explanations are possible.

Question 10: How would you test the model you proposed in your answer to question 9? (10%)

Answer 10: Your answer will clearly depend on the model you proposed in the answer to question 9. The key here is that any test of your model has appropriate controls.

Chapter 2.10: The Curse of The WERE Rabbits

Question 1: Is the size difference between WERE rabbits and East End Island rabbits primarily genetic, primarily environmental or some combination of the two? How could you test your theory? (10%)

Answer 1: In the wild, the WERE rabbits are about 80% bigger than their East End Island rabbit cousins. However, this difference essentially disappears once they are in captivity. Therefore it would appear that the difference in weight is primarily, or even exclusively, environmental. You could test this by introducing the WERE rabbits to East End Island, and the East End Island rabbits to West End Island and see whether they acquired the body size of the endemic population. If body size is primarily environmental, they will acquire the body size associated with their new environment rather than that of their original environment. Of course, whilst theoretically an ideal way of testing this hypothesis, it is ethically questionable as you would be 'polluting' the endogenous populations. In a sense the size of the captive rabbits is an appropriate test for this. An appropriate statistical test, such as a *t*-test, would demonstrate whether the weight of the captive populations of WERE rabbits and East End Island rabbits (2.4 ± 0.4 vs 2.2 ± 0.3) are really different or not. As the means and standard errors overlap, it seems unlikely that the average weight of rabbits from the two populations are statistically different.

Question 2: How are body colour and eye colour inherited in the End of the World Islands rabbits, and why do the F2 offspring occur in the ratio shown in Table 2.10.2? (15%)

Answer 2: The simplest interpretation is that body colour is inherited by a single gene where grey is dominant to golden. The eye colour is also inherited at a single gene. However, the two eye colour alleles show co-dominance or intermediate dominance where brown and red are the homozygotes and chocolate is the heterozygote.

You can explain the F2 ratios more clearly using a Punnett square. If you call the eye colour gene *Eye* where the E^B allele confers brown colour and the E^R allele confers red eye colour to produce genotypes (phenotypes) that are $E^B E^B$ (brown), $E^B E^R$ (chocolate) and $E^R E^R$ (red). Similarly let us call the body-colour gene *Body* where the B^+ allele confers grey fur and the B^- allele confers golden fur to produces genotypes (phenotypes) that are $B^+ B^+$ or $B^+ B^-$ (grey) and $B^- B^-$ (golden).

		Buck (male parent) gametes			
		$B^+ E^B$	$B^+ E^R$	$B^- E^B$	$B^- E^R$
Doe (female parent) gametes	$B^+ E^B$	$B^+B^+ E^B E^B$ grey body brown eyes	$B^+B^+ E^B E^R$ grey body chocolate eyes	$B^+B^- E^B E^B$ grey body brown eyes	$B^+B^- E^B E^R$ grey body chocolate eyes
	$B^+ E^R$	$B^+B^+ E^B E^R$ grey body chocolate eyes	$B^+B^+ E^R E^R$ grey body red eyes	$B^+B^- E^B E^R$ grey body chocolate eyes	$B^+B^- E^R E^R$ grey body red eyes
	$B^- E^B$	$B^+B^- E^B E^B$ grey body brown eyes	$B^+B^- E^B E^R$ grey body chocolate eyes	$B^-B^- E^B E^B$ golden body brown eyes	$B^-B^- E^B E^R$ golden body chocolate eyes
	$B^- E^R$	$B^+B^- E^B E^R$ grey body chocolate eyes	$B^+B^- E^R E^R$ grey body red eyes	$B^-B^- E^B E^R$ golden body chocolate eyes	$B^-B^- E^R E^R$ golden body red eyes

Question 3: What are the frequencies of the red-eye and brown-eye alleles in the mixed population of rabbits? (5%)

Answer 3: The frequencies of the two alleles are:

$$\text{Red allele} = \big((2\times 26)+33\big)/400 = 0.2125$$

$$\text{Brown allele} = \big((2\times 141)+33\big)/400 = 0.7875$$

Question 4: Is rabbit eye colour in Hardy–Weinberg equilibrium? Speculate on why or why not. (15%)

Answer 4: Let us define the frequency of the red allele as p, and the frequency of the brown allele as q. You already know (from the answer to question 3) that $p = 0.2175$ and $q = 0.7875$. Quite correctly $p + q = 1$.

The HWE allows us to predict the frequency of each genotype:
Homozygous red should be $p^2 = (0.2125)^2 = 0.045$
Heterozygous chocolate should be $2pq = 2 \times 0.2125 \times 0.7875 = 0.335$
Homozygous brown should be $q^2 = (0.7875)^2 = 0.620$
When added together these three frequencies equal 1.

The observed and predicted genotype frequencies are:

	Observed	Expected
red	0.130	0.045
chocolate	0.180	0.335
brown	0.690	0.620

You can undertake a chi-squared test to investigate whether these differences are significant. You always do a chi-squared test on the original data, the absolute numbers of individuals, which is this case is a sample of 200. Therefore multiply each 'expected' frequency by 200 to generate:

	Observed number	Expected number	$\dfrac{(O-E)^2}{E}$
red	26	9	32.11
chocolate	33	67	17.25
brown	141	124	2.33
total	200	200	51.69

Therefore chi-squared = 51.69.

For 2 degrees of freedom, the cut-off for 0.05% is 5.99, and for 0.01% is 9.21. Therefore eye colour in the rabbit population is not in Hardy–Weinberg equilibrium.

There are two key observations. There is a lack of heterozygotes and a lack of red-eye alleles. Explanations fall into two key areas:

Assortative mating, where red-eyed rabbits prefer red-eyed rabbits and brown-eyed rabbits prefer brown-eyed rabbits, would explain the lack of heterozygotes. Of course rabbits are not known for their 'choosiness' so you might seek an alternative explanation.

As there are many fewer red-eyed alleles than brown-eyed alleles in the population (remember the population started with equal frequencies of these one year ago), it is tempting to suggest that rabbits with red-eyed alleles are less fit than rabbits with brown-eyed alleles. This is entirely plausible, as you know that the WERE rabbit population is failing, but it's difficult to explain why this might specifically be caused by the red-eye colour. It also doesn't explain the lack of heterozygotes, unless you invoke a scheme where the red-eye allele is disadvantageous, and especially so in a heterozygous state.

A better answer might be to suggest that there is a second 'disadvantageous' gene/allele closely linked to the red-eye allele that confers a disadvantage. To get the lack of heterozygotes you might infer that this closely linked disadvantageous allele acts in a dominant fashion.

In reality there may be several plausible or less plausible answers!

Question 5: How would you test the hypothesis you have proposed in your answer to question 4. (10%)

Answer 5: In part this will depend on the answer to question 4. However, there are two likely issues:
Differences in mating preference, in which case undertake controlled mate choice experiments, two different females with one male and vice versa.
Differences in survival rate. Set up controlled survival experiments. The key here is that the strains must be kept under identical conditions.
Whatever experiments you propose, they must have appropriate controls and be comparative in some way or other. So 'microarrays' is a bad answer, but 'microarrays to compare relative gene expression between specific strains under specific conditions to test a specific theory' might be an excellent answer!

Question 6: How could a mutation in the rabbit *pie* cell-surface receptor gene make the host animals hypersensitive to bunnypox? Would you expect such a mutation to be dominant, recessive or intermediate in phenotype? (10%)

Answer 6: Any mutation in the rabbit *pie* gene must make it easier for the virus to enter the cell, so it must increase the quality or quantity of the cell-surface receptor. For example, this might be a *pie* overexpression mutation. In both cases, whether it's an increase in quality or quantity of the rabbit pie cell-surface receptor protein, this is a gain-of-function mutation and must be dominant, or possibly show intermediate dominance.

Question 7: How would you use the information from your discussions with Dr Burrows and Professor Gorrer to find the gene responsible for susceptibility to bunnypox? (10%)

Answer 7: The data suggest that the mutation associated with bunnypox is found almost entirely in the WERE rabbit population, but not in the East End rabbit population. Interestingly, in the mixed population that has been established for about a year, the bunnypox susceptibility gene is still predominantly in red-eyed rabbits, less apparent in chocolate-eyed rabbits and rare in brown-eyed individuals. This suggests that the red-eye mutation and the bunnypox susceptibility mutation are closely linked, meaning they are close together on the same chromosome. You could use this information to calculate the approximate position of the bunnypox susceptibility gene. These data also suggest that bunnypox susceptibility is dominant or possibly shows intermediate or co-dominance.

You also know that the bunnypox virus normally enters the cell via the cell-surface receptor encoded by the rabbit *pie* gene. Therefore you are looking for a mutation in a gene that might increase the amount of the rabbit *pie* mRNA (such as a transcription factor) or increase the pie protein's stability, quantity or activity.

Question 8: Having identified a gene that, when mutant, you suspect causes susceptibility to bunnypox in the WERE rabbits, how would you prove you'd found the correct gene? (5%)

Answer 8: The key point here is that the mutation is dominant, or at least showing co-dominance or intermediate dominance, and exerts its effect on the rabbit *pie* receptor protein or gene. Therefore, gene rescue in the mutant won't work as that requires the mutant allele to be recessive. You might start with wild-type (susceptible) individuals, add a wild-type copy of the gene and see whether these transgenic over-expressing strains are resistant, but this would only work if the original mutation is a simple over-expression gain-of-function. This may not work. A reasonable alternative would be to use RNAi to knock down the mutant gene in mutant resistant individuals and see whether they become susceptible. This would work if the mutation caused over expression of the gene or a true gain-of-function. Other answers are plausible.

Question 9: Speculate on what type of protein this new bunnypox susceptibility gene may encode and discuss whether the mutation is dominant, recessive or intermediate. (10%)

Answer 9: You are looking for a mutation in a gene that somehow increases activity of the *pie* gene or pie protein and is completely or partially dominant.

A transcriptional activator. Over-expression of the transcriptional activator means the rabbit *pie* gene is highly expressed, making excess cell-surface receptor and therefore creating more opportunities for the bunnypox virus to enter the cell. Similarly the transcriptional activator could be over-active or irrepressible.

A protein that interacts with an inhibitory ligand that blocks the cell-surface receptor. Over-expression of the protein that modifies the inhibitory ligand and prevents it from binding to the cell-surface receptor would mean that in the absence of the ligand, the cell-surface receptors are always open, therefore creating more opportunities for the bunnypox virus to enter the cell.

Again, depending on answers that have been given before, other answers are plausible.

Question 10: How are you going to answer Professor Gorrer's question? (10%)

Answer 10: The two most plausible reasons are:

The transgene has inserted into an essential gene effectively knocking-out the endogenous gene and creating a recessive lethal mutation.

Overexpression of the transgene itself is lethal.

Chapter 3.1: The Legend of *Neptune's Cutlass*

Question 1: Without resorting to sequencing the entire genome of each individual, design an experiment that would allow you to determine the lengths of the six Y-chromosome STRs. (5%)

Answer 1: You could PCR across each of the STRs and compare lengths of PCR fragments. The primers would need to anneal to the DNA immediately flanking the STR. The length of each fragment will then reflect the number of STRs at each of the six sites. Other experimental strategies would work just as well, but this is the most efficient way of doing this.

Question 2: What does the haplotype data shown in Table 3.1.1 reveal about the paternal ancestry of the Bellissimo, Guapo, Handsome and Lush families on August Bank Holiday Island? (5%)

Answer 2: Taking each family name in turn, and always trying to give the simplest plausible answer:

Handsome: all males have the same northern European Y-chromosome. This is presumably the original Handsome Y-chromosome.

Lush: 14-12-23-11-13-13 is probably the original Lush Y-chromosome as this is the most common Y-chromosome in males with the Lush family name. 14-12-23-10-13-13 has a single repeat change (11 to 10 in the 4th STR) and is probably a mutation that occurred within the Lush male lineage since they arrived on August Bank Holiday Island. 16-12-25-13-11-13 doesn't exist elsewhere within the August Bank Holiday Island families and is probably from another source, presumably from a northern European male on a passing ship. 18-13-24-10-15-16 seems to be from the Bellissimo family so demonstrates non-paternity.

Bellissimo: 18-13-24-10-15-16 is probably the original Bellissimo Y-chromosome as this is the most common Y-chromosome in males with the Belissimo family name. 18-15-24-12-16-16 seems to be the Guapo Y-chromosome, so again demonstrates non-paternity.

Guapo: 18-15-24-12-16-16 is probably the original Guapo Y-chromosome as this is the most common Y-chromosome in males with the Guapo family name. 18-13-24-10-15-16 is probably the Bellissimo Y-chromosome and demonstrates non-paternity.

Question 3: What is the probability that 12 randomly occurring injuries would all affect left eyes? (5%)

Answer 3: 2^{12}, which can be calculated more precisely as $1/4096$ or 0.00024.

Question 4: Briefly explain how asymmetrical patterns of gene expression may arise during development. (5%)

Answer 4: Asymmetrical patterns of development can arise via a variety of processes. For example, there could be a gradient of protein within a cell as mRNAs could be held at one end, so that when the protein translated it is concentrated at that end and diffuses through the cell, which creates a gradient. This could be fixed at cell division. Alternatively mRNAs could be actively transported to a specific area of a cell. There are many plausible answers to this question.

Question 5: What is the evidence that the absent left eye phenotype is genetically inherited? What is the likeliest mode of inheritance? (5%)

Answer 5: The 'absent left eye' could be explained as an autosomal dominant condition, as every affected individual has an affected parent. However, because the population is highly inbred, you need to explain clearly why it's not autosomal recessive. Affected III:4 and III:5 have an unaffected child, which demonstrates that it cannot be autosomal recessive. The most plausible explanation is that both parents are heterozygous (+/−) and affected (because the mutation is dominant) and their child has inherited the normal recessive allele from each of them (−/−). If the condition was autosomal recessive, the parents must be homozygous recessive (−/−) and all their children must also be affected (−/−). This is not the case here.

Question 6: Use a chi-squared test to determine whether there is an association between eye colour and whether or not someone is affected by ALE syndrome. (5%)

Answer 6: There are 250 individuals and remarkably there are equal numbers of individuals with blue or brown eyes (125). There are also 25 ALE-affected individuals. Therefore, if there was no relationship between eye colour and ALE syndrome, you'd expect equal numbers of affected people with blue eyes and brown eyes (25/2 = 12.5) and equal numbers of people with brown and blue eyes that are unaffected (225/2 = 112.5). The full working for the chi-square test is show here.

Note that within a statistical test it doesn't matter that you are talking about half a person!

Phenotype	Observed number	Expected number	(O − E)	(O − E)^2	$\dfrac{(O-E)^2}{E}$
brown-eye unaffected	103	112.5	-9.5	90.25	0.80
blue-eye unaffected	122	112.5	9.5	90.25	0.80
brown-eye ALE-syndrome	22	12.5	9.5	90.25	7.22
blue-eye ALE-syndrome	3	12.5	-9.5	90.25	7.22
total	250	1	1	250	16.04

As 16.04 is much higher than 3.84, significantly more brown-eyed individuals are affected by ALE-syndrome than blue-eyed individuals. The most plausible explanation is that the ALE gene and eye-colour gene are linked, meaning co-inherited, and therefore close to each other on the same chromosome. Specifically you might speculate that Captain Handsome (who had brown eyes and ALE syndrome) had one chromosome with the brown eye colour allele and a mutant copy of the ALE gene, and a second chromosome with a wild-type copy of the ALE gene and either a blue or brown eye colour allele.

Question 7: Redraw the Bellissimo pedigree shown in Fig. 3.1.7 and wherever possible annotate with individual genotypes for the *OCA2* eye colour gene. Wherever possible determine which individuals have inherited a parental chromosome or a recombinant chromosome. (10%)

Answer 7: As the ALE mutant allele is dominant, all affected individuals (dark squares and circles in the pedigree) must be heterozygous, unless both parents are affected, in which case they could be homozygous. However, in this pedigree the only offspring of two affected parents, individual IV:1, is unaffected, so all ALE-affected individuals within this pedigree are heterozygous.

As the blue eye colour is recessive, all blue-eyed individuals must have the genotype $-/-$. Any brown-eyed individual could have the genotype $+/-$ or $+/+$, but these can be distinguished if one parent or offspring is blue, in which case the brown-eyed individual in heterozygous ($+/-$), or if both parents are homozygous brown, in which case they must also be homozygous brown ($+/+$). These have been added to the pedigree here. If the genotype of a brown-eyed individual is unknown, then it is shown as $+/?$, and cannot be used for calculations of map distance, and you do not know whether any recombination has occurred.

You also know that in the first generation, individual I:1 has the brown-eye allele and the ALE mutant allele on one chromosome, and the blue-eye allele and ALE wild-type allele on the other chromosome. Taking all this together, you can determine whether any individual that is a direct descendent of I:1 and I:2 inherits a parental (P), recombinant (R) or undetermined (?) chromosome and this is marked beneath their genotype. There are 14 of these (either P or R) that can be used for calculating the recombination frequency (map distance).

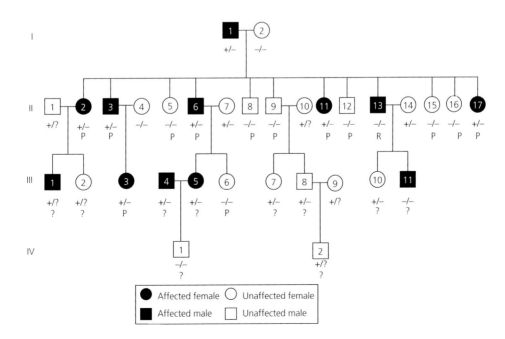

Question 8: Calculate the approximate map distance between the *OCA2* eye colour gene and the *ALE* gene. (5%)

Answer 8: There is one recombinant among 14 informative individuals. Therefore the map distance is 1/14 = 7.14 map units (see figure in question 7).

Question 9: What is the probability that Dr Townshend and islander III:3 will have a blue-eyed daughter who is affected by ALE syndrome? (5%)

Answer 9: The probability reflects chances of the child being female (0.5), having ALE from a heterozygous mother (0.5) and having a recombination event as the mother has ALE linked to the brown eye allele (1/14 = 0.0714). The probability can therefore be calculated as:

$$0.5 \times 0.5 \times 0.0714 = 0.0179 \; or \; 1.79\%.$$

Question 10: Referring to the data shown in Fig. 3.1.8, what is the most plausible genetic explanation for the ALE phenotype? (5%)

Answer 10: The ALE phenotype is most likely to be associated with a deletion. The frequency of reads of each sequence in the unaffected woman seems to be around 100 ± 20. In the affected woman most of the reads are also around 100 ± 20 but there is a long set of reads at about half the level, 50 ± 10. This suggests that this region of genomic DNA is only present at half the dose of the rest of the DNA. So in a diploid organism the simplest and most plausible explanation is that the affected sister II:17 is heterozygous for a deletion in the region where you see the 50 ± 10 reads.

Question 11: How many different *ALE* transcripts would you expect to see in affected woman II:17? What would be their approximate expected sizes and structure, and what would be their relative frequencies? (5%)

Answer 11: Each chromosome would be expected to produce a single transcript, so there will be two transcripts in total. You have no reason to believe that they are anything but equal in frequency.

The wild-type chromosome comprises all five exons (E1 to E5 inclusive) and its size will be 1591 + 810 + 903 + 782 + 1707 = 5793 bases.

The mutant is missing exon 2 so will only comprise four exons (E1 & E3 to E5) and its size would be 1591 + 903 + 782 + 1707 = 4983 bases.

So affected woman II:17 will produce both transcripts in equal frequencies.

Note also that the omitted exon 2 is 810 bases long. As 810 is a multiple of 3, its deletion will not alter the reading frame.

Question 12: Propose a model that explains why the ALE mutation in August Bank Holiday Islanders is a dominant mutation.

Answer 12: You are told that the ALE gene encodes a DNA-binding protein, so it is probably a transcription factor due to the DNA-binding domain in exon E3. You are told that individuals heterozygous for deletions of the entire ALE gene are apparently normal, so the ALE mutation is not a null and the dominance is not due to haploinsufficiency. As the mutant ALE protein is missing the protein–protein interaction domain encoded by deleted in-frame exon E2, this must be the cause of the dominant phenotype. The DNA binding domain possibly enables ALE protein to bind to DNA to switch a gene off (therefore the ALE protein functions as a transcription repressor). Perhaps a second protein normally binds to the protein–protein interaction domain to remove ALE protein from DNA binding site. In the absence of the protein–protein interaction domain, ALE protein might bind irreversibly to DNA and permanently switch a gene off. This would be a dominant mutation. (This is similar to how allolactose depresses the *lac* repressor in *E.coli.*). Note that other explanations are possible.

Question 13: How would you test the model you proposed in your answer to question 12? (5%)

Answer 13: This answer obviously depends in part on the model proposed in the answer to question 12. The key here is discussing investigations of DNA–protein or protein–protein interactions in the known wild-type and mutant ALE proteins. In addition, you could make new mutant alleles that have specific mutations in exon 2, which you might predict changes binding ability.

Question 14: What do you think the phenotype of individuals with the following genotypes would be? (10%)
(a) Heterozygous for a wild-type copy of ALE, and a mutant ALE allele with a premature STOP codon in exon E2.
(b) Homozygous for a mutant ALE allele with the same premature STOP codon in exon E2.
(c) Homozygous for a mutant ALE allele with the same missense mutation in exon E2.
(d) Heterozygous for a wild-type copy of ALE, and a mutant ALE allele with two base-pair deletion in exon E3.

Answer 14: To take each question in turn:
(a) Wild-type because of nonsense-mediated decay (NMD). The wild-type allele behaves normally, whilst the mutant mRNA is destroyed. This is effectively the same as having a deletion of one entire allele, and you already know this has a wild-type phenotype.
(b) No protein produced, so consequence unknown, as you've not encountered this before. However, it is unlikely to have an ALE phenotype.
(c) Unknown, perhaps same as (b) or nothing, or like ALE. The key point is that different missense mutations can cause different phenotypes (or at least different severities of the same phenotype) depending on the nature of specific amino acid replacements.
(d) Probably wild-type, as the mutant allele has a frame-shift mutation so it probably behaves as a null allele. However, it could be more complicated than this if the protein produced by the frame-shift mutation retains some residual function.

Question 15: Speculate on how the ALE and OGA genes and/or proteins may interact. (10%)

Answer 15: If ALE E2-deletion mutants bind irreversibly to repress gene expression, then OGA could be the protein that interacts to remove it. Therefore homozygous OGA mutants have the same phenotype as ALE E2 deletion heterozygotes. Other answers are possible.

Question 16: How could you test the model you proposed in your answer to question 15? (5%)

Answer 16: Depending on the answer to question 15, there are a number of plausible experiments that could be carried out. In the model described earlier the first approach might be to look directly for OGA ALE protein interactions in wild-type, transgenic and mutant stocks.

Chapter 3.2: The Devil's Pumpkin

This chapter is based on a University of Glasgow Genetics Finals exam from May 2011.

Question 1: Briefly explain what is meant by the term 'standard deviation'? (5%)

Answer 1: Standard deviation is a measure of the variation or spread of the data set around a mean. A small standard deviation indicates that data points are fairly tightly clustered around the mean. A large standard deviation indicates that data points are more widely dispersed around the mean.

Question 2: What, if anything, do the data in Table 3.2.1 tell you about the inheritance of cold and heat tolerance in Devil's pumpkins? (5%)

Answer 2: As these are very geographically distinct strains, they may be very different from each other. Therefore these differences (whether they are associated with cold-tolerance or other phenotypes not discussed here) are likely to be associated with variation at many genes. Nevertheless, it may be possible to find single genes with major effects on the phenotype. Here, reciprocal crosses show the same results, so there is no evidence that the 'sex' of the parental gamete (which of them is the pollinator) has an effect on the cold-tolerance phenotype. Generally the hybrids are more like the arctic strain, so there may be some alleles within the arctic strain that are dominant to those in the Mediterranean strain, but this is all speculation. The key point here it that the crosses hint at some interesting effects, but until you undertake specific controlled experiments that tell you nothing definite.

Question 3: Briefly explain how you would investigate whether the genetic basis of pumpkin pattern and colour variation in the Arctic Devil's pumpkin is the same as that for the Mediterranean Devil's pumpkin. (5%)

Answer 3: The simplest experiment would be to undertake a complementation test. For the colour gene, cross the recessive true-breeding white Mediterranean Devil's pumpkin to true-breeding white Arctic Devil's pumpkin, and see whether the F1 are white (same gene) or orange (different gene). For the pattern gene, cross the recessive true-breeding plain Mediterranean Devil's pumpkins to true-breeding plain Arctic Devil's pumpkins, and see whether the offspring are plain (same gene) or striped (different gene). Alternatively, you could take the molecular genetic approach of gene rescue. Here you take the dominant wild-type copy of the colour gene (the orange allele) from one strain and introduce it to a recessive colour gene mutant white individual from the other strain (in other words you create a transgenic strain). If the wild-type phenotype is restored then the two strains must have mutations in the same gene. However, there are potential issues with a transgenic, gene rescue approach. In particular, you may not get gene rescue (false negative result) even if the two strains are mutant for the same gene, as transgenes are not always expressed correctly in their new host. Whatever you do, try and keep the rationale behind the experiment as simple as possible.

Question 4: Using appropriate genetic notation, draw the crossing scheme that enables you to map the distance between the *plain* and *white* genes in the Arctic Devil's pumpkin. (5%)

Answer 4: Use standard notation where the alleles at the *plain* genes are P^+ (striped), which is dominant, and P^- (plain), which is recessive, and the alleles at the *white* gene are W^+ (orange), which is dominant, and W^- (white), which is recessive.

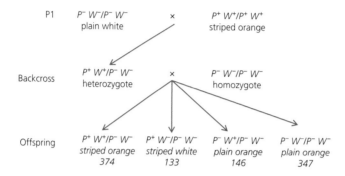

Question 5: Calculate the map distance between *plain* and *white* genes in the Arctic Devil's pumpkin. (5%)

Answer 5: Map distances are given by dividing by the number of recombinants (133 + 146) by the total number of offspring (374 + 133 + 146 + 347) and multiplying by 100 to give a percentage recombination. Therefore ((277/1000) × 100) = 27.7. The map distance in the Arctic Devil's pumpkin (27.7) is very similar to the published data in the Mediterranean Devil's pumpkin (28.0).

Question 6: Calculate the map distance between the *plain* and *white* genes in the cross between the Arctic and Mediterranean strains. (5%)

Answer 6: Map distances are given by dividing by the number of recombinants (21 + 23) by the total number of offspring (475 + 21 + 23 + 481) and multiplying by 100 to give a percentage recombination. Therefore ((44/1000) × 100) = 4.4, which is much smaller than the recombination distances (28.0 and 27.7) found in identical crosses within the Mediterranean and Arctic strains.

Question 7: Speculate on why the map distances between the *plain* and *white* genes in the Arctic/Mediterranean F1 hybrid is significantly different from the map distance between the *plain* and *white* genes in the original pure-breeding Arctic and Mediterranean strains? (5%)

Answer 7: You know the following information about the distance between the *plain* and *white* genes:

Mediterranean × Mediterranean distance = 28 cM
Arctic × Arctic distance = 27.7 cM

Which implies that within these strains they are probably in the same place on the same chromosome. However:

$$\text{Arctic} \times \text{Mediterranean distance} = 4.4\,\text{cM}$$

This could be explained by an inversion of around 23 cM in length between (but NOT including) the *plain* and *white* genes in one of the strains. The genes really are still about 28.0 map units apart, but any recombination occurring within the inversion will result in unbalanced chromosomes, and therefore mean that any resulting eggs or sperm are inviable. So it will seem as if very few recombination events have taken place.

Question 8: How would you identify the DNA sequence that is mutated by insertion of the modified T-DNA? (5%)

Answer 8: As you know the sequence of the T-DNA, what you are looking for is the host DNA sequence that is adjacent to the inserted T-DNA. Historically this would have been achieved via transposon tagging, but this is now often done by simply sequencing the entire genome.

Question 9: Explain the function of each of the three components of the transgene:
(a) the T-DNA ends,
(b) the *kanamycin resistance* gene, and
(c) the five directional Devil's pumpkin enhancer elements. (5%)

Answer 9: Taking each component in turn:
(a) The T-DNA ends enable random integration into the host chromosome.
(b) The kanamycin resistance gene acts as a selectable marker. Only Km^R plants have taken up the transgene.
(c) The five directional Devil's pumpkin enhancer elements act as a dominant up-regulator of gene(s) immediately downstream of the insertion site.

Question 10: Speculate on why the CT1 insertion in the Arctic Devil's pumpkin causes a dominant cold-tolerant phenotype. (5%)

Answer 10: CT1 is an insertion upstream of the gene, which may upregulate the *Frozen* gene. This would require the enhancers within the inserted T-DNA to be facing towards the promotor (facing to the right in figure 9). Over-expression of a gene often gives a dominant gain-of-function phenotype. Other answers are possible.

Question 11: Speculate on why each of the three insertions (CS1, CS2 & CS3) in the Arctic Devil's pumpkin each cause recessive cold-sensitive phenotypes. (5%)

Answer 11: Taking each of the three CS insertions in turn:
CS1 disrupts the control region and may therefore physically interrupt, or at least inhibit, a sequence that helps direct expression of *Frozen*. Therefore you will have less or no expression

of the gene. (It may be that in this case the enhancers face to the left, but that is not necessarily the case). This is a loss-of-function mutation and is likely to be recessive.

CS2 & CS3 are insertions that interrupt the coding sequence, almost certainly making the protein encoded by *Frozen* non-functional. Again this is a loss-of-function mutation and is likely to be recessive.

Question 12: Explain what you would expect the phenotype to be of:
(a) a CT1/CS1 heterozygous plant grown in the cold?
(b) a CT1/CS2 heterozygous plant grown in the cold? (5%)

Answer 12: Both are cold tolerant as the dominant over-expressed CT1 allele is very likely to mask the effect of the recessive under-expressed or not expressed or non-functional CS1 and CS2 alleles.

Question 13: Explain what you would expect the phenotype to be of:
(a) a CT1 CS1 homozygous plant *Fro* (CT1 CS1) / *Fro* (CT1 CS1)
(b) a CT1 CS1 heterozygote plant *Fro* (CT1 CS1) / *Fro* (+). (5%)

Answer 13: Taking each combination in turn:
(a) This could be cold sensitive as CS1 (loss of function) within the control region may well mask the CT1 (gain of enhancer) within the same of control region. However, the reverse is possible.
(b) This would depend on the answer to (a). If the double mutant allele has a CS1 phenotype, then this loss-of-function mutation will be recessive to wild-type. However, If the double mutant allele has a CT1 phenotype, then this gain-of-function mutation is likely to be dominant to wild-type.

Question 14: Using the data in Table 3.2.4, speculate on the function of the *Frozen* gene. How would you test your hypothesis? (10%)

Answer 14: Five of the eight upregulated genes (*Anna*, *Olaf*, *Hans*, *Bulda* and *Pabbie*) seem to be dependent on *Frozen* expression. However, the expression of three genes (*Elsa*, *Kristoff* and *Marshmallow*) seem to be independent of *Frozen* expression. You may speculate that *Frozen* encodes a transcription factor that upregulates *Anna*, *Olaf*, *Hans*, *Bulda* and *Pabbie*. Other answer may also be possible.

To test this you might investigate whether the Frozen protein can bind upstream of the *Anna*, *Olaf*, *Hans*, *Bulda* and *Pabbie* genes.

Question 15: What does the data in Table 3.2.5 tell you about the genetic basis of drought tolerance in Arctic Devil's pumpkin? (5%)

Answer 15: As none of the genes show significant differences in expression between wild-type and *Frozen*(CS1) strains, it suggests that *Frozen* is not in any way linked to any response to drought.

Note that *Elsa* is upregulated in both environments, even in *Frozen*(CS1) mutants, so it is perhaps a general stress response gene independent of *Frozen*.

Also note that *Hans* and *Pabbie* are dependent on *Frozen* for their expression in the cold, but independent of *Frozen* in drought conditions, so the two genes may have a different mode of expression in the two environments.

Question 16: Speculate, giving your reasoning, on which of the eight base-pair differences found between the Arctic and Mediterranean Devil's pumpkin could be the cause of the difference in cold tolerance known to be associated with the *Frozen* gene. How would you test your hypothesis? (10%)

Answer 16: Taking each difference in turn and remembering that gain-of-function *Frozen* mutations lead to cold tolerance:

The DNA sequence changes at −502 (a potential enhancer mutation) and at −10 (a potential promoter mutation) could result in upregulation of *Frozen* in cold environments.

The DNA sequence changes at codon 9 and codon 201 are silent as they do not change the amino acid encoded, so it is difficult to see how they affect *Frozen* protein function.

The DNA sequence changes in codon 69, codon 101 and codon 175 cause missense mutations where one amino acid is substituted for another. To get cold sensitivity by this route the protein would have to acquire higher function, activity or perhaps stability. You have no idea, from the limited data given, whether this could be the case or not. But it remains a possibility.

The DNA sequence change in the 3′ UTR of the mRNA could increase *Frozen* mRNA stability and result in more Frozen protein being produced.

Remember that all eight differences exist between the Arctic and Mediterranean strains. Each of these could be investigated by creating eight transgenic strains, each of which only carries one of the mutations. You might also want to make reporter constructs with specific *Frozen* mutations (where mRNA levels are implicated), or look directly at protein function or stability (where protein function or stability is implicated)

Question 17: Describe in outline a series of experiments you would undertake to productively investigate drought tolerance in the Irish Devil's pumpkin. You might wish to refer to data from previous questions. (10%)

Answer 17: Here it would be sensible to refer to the drought tolerance genomic data shown previously, especially the fact that three genes (*Elsa*, *Hans* and *Pabbie*) have already been implicated in drought tolerance in WT and *Frozen*(CS1) transcriptomics. There are many possible experiments that could be undertaken, but the key here is to suggest a clear experiment that has well-defined aims and appropriate controls.

Chapter 3.3: Gravity

A version of the questions in this chapter first appeared in the Genetics finals examinations at the University of Glasgow in 2010.

Question 1: Calculate the mean exit scores of wild-type flies when the maze is held vertically (Fig. 3.3.4) or horizontally (Fig. 3.3.5). What does this tell you about the gravitactic behaviour of wild-type flies in a maze? (5%)

Answer 1: Calculate a simple mean from the data in Fig. 3.3.4.

$$\frac{(2\times4)+(5\times6)+(7\times7)+(11\times8)+(25\times9)}{2+5+7+11+25} = \frac{400}{50} = 8.0$$

Calculate a simple mean from the data in Fig. 3.3.5.

$$\frac{(1\times1)+(2\times2)+(3\times3)+(10\times4)+(18\times5)+(8\times6)+(6\times7)+(2\times8)}{1+2+3+10+18+8+6+2} = \frac{250}{50} = 5.0$$

Question 2: Using appropriate notation, draw the cross described. Show clearly how the newly mutated genes are passed through the generations and explain the rationale behind the crossing scheme, and in particular why you only screen the F2 males (and not their sisters) for abnormal gravitactic behaviour. (5%)

Answer 2: The figure below shows the inheritance of X- and Y-chromosomes in wild-type females (XX) and wild-type 'mutagenised' males (X*Y*) where the asterisk (*) indicates the chromosome carries a number of as yet unidentified mutations.

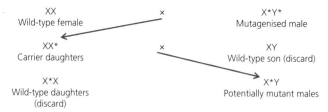

The mutagenised males eat the mutagen in their food, which has the potential to damage any DNA in any cell. Here the objective is to mutagenise sperm which will carry mutations on all chromosomes. X-linked mutant sperm will pass from father to daughter (who will be heterozygous) and then to both sons and daughters in the F2.

X-linked recessive mutations will be seen in 50% of the F2 males, but all recessive mutations in their F2 sisters will be masked as they are all in a heterozygous state. Therefore in males the phenotype reflects the genotype (as there is one copy of each X-linked gene) but in females it will not (as there are two copies of each X-linked gene).

Question 3: How would you demonstrate that the abnormal gravitactic phenotype in each *apollo* strain is caused by an X-linked recessive mutation? (5%)

Answer 3: You could set up a series of controlled crosses. If a mutation is X-linked, a male cannot pass it to its sons. So the key cross is pairing a homozygous wild-type female (+/+) with a mutant male (−/Y), in which case the daughters will all be carriers (+/−) and the sons will all be wild-type (+/Y). The F1 sons cannot pass the mutation on to their F2 offspring, as they do not carry it. The F1 daughters will have sons in the ratio of 1:1 affected vs unaffected, irrespective of who she mates with, as her sons get their X- chromosome from her.

Question 4: Having established that all 12 *apollo* mutations are on the X-chromosome, how would you determine whether they are mutations in the same or different genes? (5%)

Answer 4: You would undertake a complementation test. Specifically you would cross the independently derived mutants to each other and look at the phenotype of their offspring.

  If the parents have mutations in the same gene, then the mutations fail to complement and all the offspring will have a mutant phenotype.

  If the parents have mutations in different genes, then the mutations complement and all the offspring will have a wild-type phenotype.

Question 5: Draw the entire crossing scheme for the three-point test cross. Clearly show all the genotypes and phenotypes and explain the rationale behind the crossing scheme. (10%)

Answer 5: The full crossing scheme is:

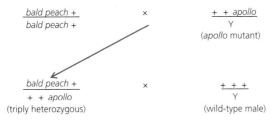

The F2 males would have any one of eight possible genotypes (as shown in Table 3.3.1 in the original question). As you are mapping genes on the X-chromosome and only looking at males, the genotype will reflect the phenotype.

Question 6: Calculate the relative map distances between *apollo*, *bald* and *peach*. Draw the resulting genetic map. (10%)

Answer 6: Taking each pair of genes in turn:

$$apollo \text{ vs } bald = \left(\left(10 + 1675 + 1590 + 5\right) / 10000\right) \times 100 = 32.80$$

$$apollo \text{ vs } peach = \left(\left(10 + 25 + 30 + 5\right) / 10000\right) \times 100 = 0.70$$

$$bald \text{ vs } peach = \left(\left(1675 + 25 + 30 + 1590\right) / 10000\right) \times 100 = 33.20$$

A topologically correct map, approximately drawn to scale, would look like:

```
bald -----------------------------32.80 ----------------------------apollo-0.7-peach
     -------------------------------------------------------33.20----------------------------------
```

Note that it can be drawn the opposite way round (*peach, apollo, bald*).

Note that longer distances are always underestimates, as double crossovers are not included in the calculation.

Question 7: Briefly describe how the fact that *soyuz* is associated with a *P*-element insertion makes cloning the mutant gene relatively easy. (5%)

Answer 7: As the *P*-element DNA sequence is known, it acts as a 'molecular tag'. Therefore if you can identify the endogenous DNA sequence adjacent to the *P*-element insertion site, you can use data from the fly genome project to identify the precise DNA sequence into which the *P*-element has inserted.

Question 8: How would you prove that the gravitactic phenotype in *soyuz* was caused by the *P*-element insertion and not by the missense (S456T) mutation? How is it possible for a missense mutation to cause no phenotype? (5%)

Answer 8: You'll need to create a mutant that only has one of the two mutations and look at its phenotype. This can be achieved in one of at least three ways.

It may be formally possible to split the two mutations apart by recombination, but they are so close together that any crossing scheme would have to be attempted on an industrial scale; so, this is a plausible but unrealistic option.

There are ways of jumping the *P*-element out of its insertion site by crossing the mutation strain to one carrying a constitutively active transposase gene. If you can identify flies that have lost the *P*-element you can look at their phenotype and see whether the remaining S456T missense mutation alone results in a mutant or wild-type phenotype. One potential problem here is that *P*-element excision doesn't always happen precisely, so whilst the *P*-element may excise, it may leave behind several base-pair changes, a small deletion, or a small insertion.

It may also be possible to use genome-editting technology such as CRISPR to revert the S456T missense mutation in the fly genome. An alternative might be to generate a transgene containing the S456T mutation alone. But to determine whether it causes the phenotype, it would have to be expressed in a null mutant of the gene (as you know it is recessive). If it rescued a null mutant and had no other phenotype, this would show it was not the cause of the gravitactic phenotype.

In a missense mutation, one amino acid is replaced by another. This may have no effect on protein function because (a) the two amino acids may have similar biochemical properties and/or (b) the amino acid in question is not in a critical functional domain of the protein or (c) some combination of the two.

Question 9: Using the information in Fig. 3.3.7 & Table 3.3.2, speculate on why males from each of the 12 strains show the indicated phenotype. (10%)

Answer 9: Taking each mutation in turn:

apollo¹	Point mutation either premature stop or critical missense. Presumably a null, non-functional protein.
apollo⁶	Point mutation missense involving less critical amino acid Protein affected but has some function.
apollo⁹	Point mutation. Probably promoter mutation probably reducing expression.
P1	Insertion upstream of gene having no effect.
P2	Insertion downstream of gene having no effect.
P3	Insertion disrupts both transcripts, null mutation (implies it's an essential gene).
soyuz (P)	Insertion in intron disrupts splicing from upstream site trans-splicing from exon 1 into *P*-element and terminates. No protein from promoter 1, doesn't affect promoter 2.
Deletion A	Deletes most of the gene (gene is essential).
Deletion B	Deletes first exon from promoter 1 (implies transcript 1 is needed for gravitaxis).
Deletion C	Deletes second exon from promoter 2 (confirms transcript 2 is essential).
Deletion D	Deletes first promoter (non-lethal) and something essential upstream.
Deletion E	Deletes small bit of (essential) exon 3, so must be in-frame.

Question 10: Speculate on why males carrying inversion 1 are slow moving and have no gravitactic behaviour. Similarly speculate on why females that are heterozygous for inversion 1 are slow-moving but have normal gravitactic behaviour. (5%)

Answer 10: Taking each part of the question in turn:

Males carrying inversion 1 are slow moving and have no gravitactic behaviour because the inversion affects promoter 1, effectively meaning the males cannot make the gravitactic transcript. As there is a secondary movement phenotype that seems to be unrelated to the gravitactic gene, the inversion may affect expression of an upstream gene that affects the amount of movement.

Females that are heterozygous for inversion 1 are slow-moving, but have normal gravitactic behaviour because the wild-type allele of promoter 1 drives the gravitactic gene normally so they show normal gravitactic behaviour. However, if the inverted promoter 1 now drives expression of an upstream gene associated with movement, this 'misexpression' is likely to manifest in a dominant fashion.

Question 11: Draw a cartoon of the intron/exon structure of the *apollo* gene based on the information in Fig. 3.3.7 showing which sequence you will use as a probe for northern analysis to detect all *apollo* transcripts. In addition, indicate sequences that you could use as probes to detect each *apollo* transcript individually. (5%)

Answer 11: This is illustrated in the figure here:

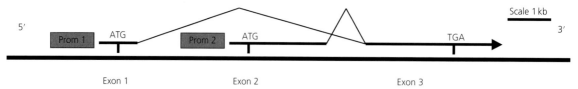

To detect all transcripts use a sequence from exon 3.
To detect the gravitactic transcript use a sequence from exon 1.
To detect the essential transcript use a sequence from exon 2.

Question 12: Speculate on how the EMS-induced mutations in hemizygous *apollo¹* and *apollo⁶* mutant males result in the patterns of bands shown in Fig. 3.3.8. (5%)

Answer 12: Taking each transcript and mutation in turn:
The gravitactic transcript is the small 2.2 kb one. It is only expressed in the head.
The essential transcript is the large 3.6 kb one and is expressed everywhere.
The *apollo¹* mutant has no gravitactic transcript. As it is a point mutation within the coding region you might speculate that this is due to the presence of stop codon that results in non-sense mediated decay.
The *apollo⁶* mutant makes a normal transcript so must be a missense mutation.

Question 13: What is the most likely explanation for the banding pattern seen in the *soyuz* hemizygous mutant males in Fig. 3.3.8 & Fig. 3.3.9? How would you test your hypothesis? (5%)

Answer 13: According to Fig. 3.3.8, the *soyuz* mutant has no gravitactic transcript but a normal essential one; however, Fig. 3.3.9 reveals that it has a small gravitactic transcript. The most plausible explanation is that promotor 1 functions normally but exon 1 splices into the *P*-element and the mRNA truncates there. You could test this by sequencing the transcript.

Question 14: Speculate on why *apollo⁶/ariane* females have a wild-type phenotype. How would you test your hypothesis? (5%)

Answer 14: This must be intragenic complementation. The two missense mutated amino acids presumably interact at a protein level and effectively cancel each other out. You could test this by investigating which amino acids within the Apollo protein interact with each other.

Question 15: In light of all the information here, propose a model to explain the function of the *apollo* gene. What experiments would you undertake to test your model? (15%)

Answer 15: There are any number of plausible experiments that could be attempted here. The key is to design, with appropriate controls, experiments that use the stocks already available and the array of genetic techniques available to answer questions about gravitaxis that add something to the body of knowledge already determined.

Chapter 3.4: Kate & William: a love story

A version of the questions i this chapter originally appeared as a 3-hour examination for the final year Genetics students at University of Glasgow in 2012.

Question 1: January 2013 was particularly cold, with frequent deep snow recorded across Scotland, especially around St Andrews (see Fig. 3.4.4). In light of this speculate on why there may have been such a large fall in the frequency of brown haggis between June 2012 and June 2013. How would you test your theory? (10%)

Answer 1: As a rule, in any answer where you're asked to speculate, the simplest plausible answer is probably the best answer. Here you have cold snowy winters where brown haggis numbers are decreasing dramatically, but white haggis numbers are not. Therefore the most plausible answer is that in snowy conditions the brown haggis are more obvious than the whites, and are therefore eaten by the local (unnamed) predator. Therefore an answer invoking camouflage is good. This could be tested via field observations or perhaps via laboratory experiments with an appropriate predator (although there may be ethical issues with this approach).

There are alternative, and ever more complex, possibilities. For example perhaps brown fur is less cold tolerant than white fur, but this will need more explanation as it's not immediately obvious why that should be (though it's a long way from being a wrong answer). Therefore, in questions like this there are likely to be several possible (and non-overlapping answers). This could be tested by looking at cold-tolerance in white and brown haggis.

Question 2: What is the evidence that the white body-colour gene is X-linked? (5%)

Answer 2: The key point is that there is no male to male transmission.
Note: The answers to questions 3–5 are best understood by annotating the original Fig. 3.4.2. In the figure below the chromosomes coloured in red must carry the mutant *white* allele.

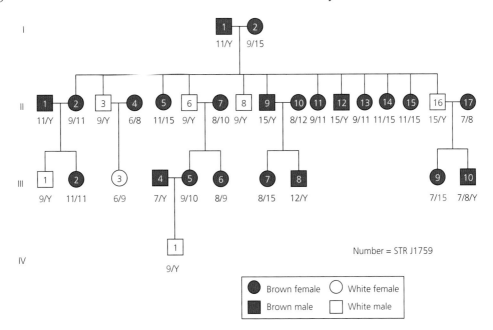

Question 3: The evidence from Fig. 3.4.6 clearly demonstrates that the white body colour mutation and the STR are closely linked on the X-chromosome. Which individual(s) show evidence of recombination? (10%)

Answer 3: The great grandfather (I:1) is brown. You have already established that it is an X-linked condition that must come from the brown great-grandmother (I:2) and can be seen in her sons (generation II). She has six sons, three of whom have inherited the 9 STR allele and are white, and three of whom have inherited the 15 STR allele (two brown and one white). The most plausible explanation is that 9 and white are linked. Therefore individual II:16 (a white male with genotype 15/Y) must be a recombinant. It is not necessary to invoke any other recombination events. Note that as clearly stated in the preamble and question, the STR is not the cause of the phenotype. Therefore the most plausible explanation is that a novel X-linked *white* mutation has occurred near to a 9 STR.

Question 4: Speculate why individual III:3 is white. (5%)

Answer 4: Female II:4 must be heterozygous and carrying a *white* allele at the same colour gene as the rest of the haggis pedigree. The *white* allele is probably segregating with STR 6 in that individual.

Question 5: If individual III:6 is mated to an unrelated brown male with a STR genotype 14/Y, what is the probability that their firstborn pup will be a white son with STR genotype 8/Y? (5%)

Answer 5: Here you are crossing a mother, III:6, who has the genotype (8-brown/9-white) with a father who has the genotype (14-brown/Y), to generate a son who is (8-white/Y).
Probability of getting the Y-chromosome from the father is = 0.5.
Probability of getting the *white* allele from the mother is = 0.5.
Probability of getting white and 8 from the mother requires a recombination event and the preamble in the text tells us the recombination frequency = 0.15.
Multiply together $0.5 \times 0.5 \times 0.15 = 0.0375$ (or 3.75% if you prefer).

Question 6: What is the genetic basis of 'footedness' in haggis? (5%)

Answer 6: There is a lot of information given which generates a very simple answer. It is an X-linked gene where the *Left* allele is dominant to the *Right* allele.

Question 7: Speculate on how the C to T change in the *chirality* (HG43) gene affects 'footedness' in haggis? (5%)

Answer 7: If *Left* (CGA: arginine) is dominant to *Right* (UGA: STOP), then the premature stop codon in *Right* is likely to lead to a non-functional truncated protein. The simplest explanation is that *Right* is a recessive loss-of-function (LOF) mutation. It follows that *Left* (which has DNA-binding ability) probably encodes a transcription factor that enhances/promotes long left legs. In the absence of a functional *Left* allele, a haggis will have long right legs.

Question 8: Calculate the mean (average) survival in days for each of the four classes of haggis shown in Table 3.4.1. (5%)

Answer 8: Taking each class in turn:

Brown female	5.1
Brown male	6.0
White female	45.1
White male	45.0

Question 9: What statistical test should you use to prove that white haggis survive the *Bacillus catastrophicus* infection better than brown haggis? Why is this the appropriate statistical test? (5%)

Answer 9: Analysis of variance (ANOVA) or *t*-test.

Question 10: Why are white haggis always relatively resistant to *Bacillus catastrophicus* infection, and why are brown haggis always relatively sensitive to *Bacillus catastrophicus* infection? (5%)

Answer 10: All white males have a mutant T in common, which is always absent in the brown males. This is in the coding region of the *white* gene so presumably causes the LOF white mutation. The mutant T is also just upstream of the transcription start site of the immune system gene *wallace*, so resistance is probably associated with a change in the transcriptional control of *wallace*. So the T presumably alters a control site upstream of *wallace* (possibly a promoter or enhancer site). Therefore a single base-pair change affects two genes: the coding region of *white*, and the control region of *wallace*.

Question 11: Speculate on the evolution of *Bacillus catastrophicus* resistance in haggis. (5%)

Answer 11: In the region shown, the DNA sequence of brown males is very variable, whereas the DNA sequence of white males is identical. This suggests the *white* mutation has arisen recently and is associated with a selective sweep through the population.

Question 12: What does the qPCR analysis tell you about the genetic basis of *Bacillus catastrophicus* resistance in Scottish haggis? (5%)

Answer 12: The mRNA level of *wallace* is higher in white males relative to brown males. Therefore resistance is probably associated with the amount of *wallace* mRNA/protein. The *wallace* resistant mutation is therefore an overexpression mutation, again suggesting that the T change described earlier affects the promoter or another aspect of the *wallace* control region. In an answer it would be sensible to state explicitly that qPCR measures mRNA levels.

Question 13: If you exposed a heterozygous brown female to *Bacillus catastrophicus* infection, would you expect her to be resistant, susceptible or somewhat intermediate? Explain. (5%)

Answer 13: Overexpression mutations are usually dominant. Therefore these heterozygous brown females are probably more resistant than their homozygous brown sisters. Note that you can also make a strong argument for them being intermediate between homozygous brown and homozygous white if resistance is simply a product of mRNA levels.

Question 14: Taking each of the ten genes in turn, explain why each gene has the expression levels in homozygous brown females, homozygous white females and heterozygous females shown in Table 3.4.2. (10%)

Answer 14: Taking each gene in turn:

white Expressed slightly higher in brown/brown relative to brown/white. Not expressed in white/white. A simple single gene brown dominant to white as discussed earlier.

wallace Highly expressed in white/white and brown/white. Note that the confident limits overlap so suggests that the answer to question 14 is that brown/white females are just as resistant as their white/white sisters.

bac1, bac2, bac3 and *fun2* Immune system genes that reflect *wallace* expression levels and may even therefore be under transcriptional control of *wallace*.

bac4, fun1, fun3 and *fun4* Immune system genes that are equally expressed across all genotypes and are therefore probably not under the control of *wallace*.

Note that it's only the relative gene expression within a row that is important, not the absolute gene expression. Using *bac1* and *bac2* as an example the pattern is identical (low high high in the three columns), and the fact that the actual numbers are different (*bac2* is always more highly expressed than *bac1*) is relevant. In addition remember to take account of confidence limits as (for example) 83 ± 16 and 69 ± 14 overlap so are unlikely to be significantly different.

Question 15: Briefly describe a model for how mutations in the *wallace* gene may cause resistance to *Bacillus catastrophicus* in Scottish haggis. (5%)

Answer 15: The single base mutation to a T in the control region of the immune system gene *wallace* increases the level of expression of the DNA-binding protein *wallace*. This is the same single base-pair change that causes a LOF mutation in *white* (mutant phenotype from brown to white). Over-expression of *wallace* leads to over-expression of downstream genes *bac1, bac2, bac3* and *fun2*, one or more of which mediate the response to *B. catastrophicus*.

Question 16: You have enough funding to make genetically modified haggis at the University of St Andrews haggis transgenic facility run by Professor Paula Bear. Design three recognizably different *wallace* transgenes that might make genetically modified haggis resistant to *Bacillus catastrophicus* and explain how each transgene works. (10%)

Answer 16: Three transgenes could include:

Over-expression – *wallace* under the control of a strong promoter, possibly the mutant one.

mRNA stability – *wallace* with some modification increasing mRNA stability.

bac1, bac2, bac3 or *fun2* — overexpression.

DNA-binding modification – site-directed mutagenesis of *wallace* DNA-binding motif.

There are many subtly different transgenes that could contribute to this answer.

Chapter 3.5: The Titanians

Question 1: When you have acquired multiple random DNA sequences from the Titanians, explain how you could build a full, ordered genome. What problems might you encounter and how might you resolve them? (5%)

Answer 1: In theory building a full ordered genome sequence is simply a matter of collecting lots of short DNA sequences from the host organism, searching for homology and placing sequences that share homology next to each other. However, building an entire *de novo* genome is much more complicated than this. Some host DNA sequences may be refractive to the sequencing technology employed, so will never be sequenced and will therefore be 'missing' from the genomes sequence. Using an alternative sequencing technology may resolve this. More problematic is the fact that many genomes contain repetitive sequences, and as these may exist at different sites in the genome they are going to confuse any strategy that uses 'head-to-tail' homology as the main route by which ordered genomes are constructed. Genomes are therefore usually constructed by mapping DNA sequences onto well-established landmarks. This could be a gene sequence that has been mapped to a specific position on a specific chromosome, mapped chromosomal abnormalities (mapping breakpoints) and so on. In this case you don't know how many 'well-established landmarks' within the Titanian genome are actually known to us.

Question 2: What does the data in table 1 tell you about the inheritance of antennal fluorescence colour in the different Titanian populations? (5%)

Answer 2: As you are led to believe that the methane environment within which each population is living in the CHEAT facility is identical, you can conclude that the colour of antennal fluorescence is genetic in origin. Furthermore, the data suggest that each population is a true-breeding strain, so that at least as far as genes associated with fluorescing antennal colour are concerned, each population is homozygous for fluorescence antennal colour genes. It's also clear that as far as fluorescence antennal colour is concerned the populations are genetically distinct.

Question 3: Propose a model that explains the genetic basis of fluorescing red and yellow antennal colour in Titanians (10%)

Answer 3: The fact that the sons and daughters have different phenotypes suggests that either the gene is on the sex-chromosomes, or the gene is expressed in a sex-specific fashion. As it was flagged earlier that the Titanians appear to have sex-chromosomes so proposing a mode of inheritance of X-linked genes is probably the best place to start. You also know that the parents are homozygous for this gene as they come from true-breeding populations.

Looking at the data you see that sons have the same phenotype as their mothers, irrespective of the father's phenotype. This is similar to the behaviour of mammalian X-linked genes where females are XX and males are XY. It follows that if the mother is homozygous for an X-linked gene allele (XX), hers sons (XY) must be hemizygous for the same allele, and exhibit the same phenotype.

Following the same theme of X-linkage, all the daughters would have to be heterozygous (red/yellow), which at first you might think would generate orange daughters, but it doesn't, as the daughters exhibit a range of colours from yellow to red. Again following the theme of how mammals deal with X-linked genes in females this range of colours could reflect some consequence of dosage compensation, where each cell randomly switches off a one X (so male and female cells then only use one X). Let us suppose the colour producing cells in a heterozygote randomly switch of an X-chromosome. Then in a heterozygous female some colour producing cells will express the red allele and others the yellow allele. As a result statistically you would expect to see a bell curve of colours, where yellow (all cells expressing the yellow allele) and red (all cells expressing the red allele) is far less likely than roughly half the cells expressing red, and the other half expressing yellow. This is exactly what you see.

Question 4: Propose a model that explains the spread of colours in the daughters seen in figure 3.5.6. (5%)

Answer 4: Whilst the F1 'various' mothers have a wide range of colour phenotypes from yellow to red, genetically they are all identical, being heterozygous for yellow/red. Therefore here you are simply crossing heterozygous mothers to hemizygous fathers. As it's an X-linked gene, the phenotypes of the sons simply reflect the genotypes of their mother. As the mothers are heterozygous yellow/red, their sons will be 50% yellow and 50% red.

When you mate heterozygous (yellow/red) mothers to a red father (figure 3.5.6a) half their daughters will be homozygous red and the other half will be heterozygous (yellow/red) showing a spread in bell-curve shape from yellow to red. This is exactly what you see in figure 3.5.6a. The same female mated to a yellow male will, using the same argument, have the spread of daughters shown in figure 3.5.6b.

Question 5: Speculate on what kind of protein(s) the colour gene(s) may encode and how this causes the spread of colours seen. How would you test your hypothesis? (5%)

Answer 5: You have been told that the fluorescing pigment is made in a cluster of cells above the brain between the bases of the antenna and transported up the length of the antenna where it fluoresces. This processes requires the synthesis of fluorescing pigment (in Titanians this pigment could be a protein and therefore encoded by a gene directly, or if it's not a protein then it may synthesised by enzymes in a biosynthetic pathway and the gene could therefore encode an enzyme). In addition this fluorescing pigment needs to be carried from the synthesising cells at the base of the antenna to the tip of the antenna, so some kind of carrier protein may be required. You could sequence the gene and see whether it has homology to other known fluorescing pigment, enzyme or carrier proteins (though you have been told it has no DNA sequence homology with any terrestrial species). Alternatively you could express the gene under controlled conditions and investigate the function of the protein the gene synthesises.

Question 6: If, as Professor Celeste Teal-Boddie suggests, the light pulses are an essential component of the red-fluorescing Titanians courtship ritual, what message might they be trying to get across? (5%) 51

Answer 6: In Morse code the light pulses reveal the message 'S' (dot dot dot), 'E' (dot), 'X' (dash dot dot dash), 'Y' (dash dot dash dash)

Question 7: In generation 15, what are the frequencies of the blue and green alleles?

Answer 7: In generation 15 the frequency of each colour type is said to be blue 51%, blue/green 29% and green 20%. Taking each allele in turn:
The frequency of the blue allele must be $((2\times51)+29)/2 = 65.5\%$ or 0.655
The frequency of the green allele must be $((2\times20)+29)/2 = 34.5\%$ or 0.345

Question 8: In generation 15, is the Titanian population in Hardy-Weinberg Equilibrium? (5%)

Answer 8: Let us define the frequency of the blue allele as p, and the frequency of the green allele as q. You already know (from the answer to question 7) that $p = 0.655$ and $q = 0.345$. There are only two alternative alleles at this gene, so quite correctly $p+q =1$.
The HWE equation allows us to predict the frequency of each genotype:
Homozygous blue should be $p^2 = (0.655)^2 = 0.429$
Heterozygous blue/green should be $2pq = 2 \times 0.655 \times 0.345 = 0.452$
Homozygous green should be $q^2 = (0.345)^2 = 0.119$
When added together these three frequencies equal 1.
The observed and predicted genotype frequencies in generation 15 are therefore:

	Observed	Expected
blue	0.51	0.43
blue/green	0.29	0.45
green	0.20	0.12

You can undertake a chi-squared test to investigate whether these differences are significant. You always do a chi-squared test on the original data, the absolute numbers of individuals, which is this case is a sample of 100. Therefore:

	Observed number	Expected number	$\dfrac{(O-E)^2}{E}$
blue	51	43	1.49
blue/green	29	45	5.69
green	20	12	5.33
TOTAL	100	100	12.51

Therefore chi-squared = 12.51.
For 2 degrees of freedom, the cut-off for 0.05% is 5.99, and for 0.01% is 9.21. Therefore in generation 15 the mixed blue/green population is not in Hardy-Weinberg Equilibrium.

Question 9: What, if anything, could the frequencies of blue, green and blue/green Titanians in generation 15 tell you about the mating behaviours of blue and green Titanians? Are there any alternative explanations, and how would you test your theory? (10%)

Answer 9: The changes in allele frequencies, coupled with the fact that the mixed blue/green population is not in Hardy-Weinberg equilibrium, suggest that some kind of selection is happening, or it may be that the population is exhibiting non-random mating. In fact there is an excess of homozygous individuals relative to heterozygous individuals. This may be because the heterozygotes are less fit (which could be tested by survival assays) or less fertile (which could be tested by fertility assays). Alternatively there may be assortative mating in which blue males are more likely to mate with blue females, green males with green females and blue/green males with blue/green females (which could be tested by female choice or male competition assays).

Question 10: Draw an annotated crossing scheme for the mating between a *bald* female and wild-type male that supports Professor Pluto's hypothesis that the Confused-1 and Confused-2 strains have a mutation in an autosomal recessive sex-determining gene. (10%)

Answer 10: You are told that the *bald* mutation is on the X-chromosome and for the purposes of illustrating this cross let's call the wild-type *bald* allele b^+ and the mutant *bald* allele b^-.

Let's call the wild-type allele of the autosomal sex-determining gene s^+ and the mutant allele s^-. You are told that males homozygous for this sex-determining mutation (s^-/s^-) are sterile, but at this point have no idea whether the same is true in females. But at the very least the fathers in the cross cannot be homozygous for the sex determining mutation (because they'll produce no offspring). Let's work on the hypothesis that both parents are heterozygous for the autosomal recessive sex-determining mutation (meaning they are both s^+/s^-).

Therefore the original cross is:

$$\text{female } (b^-/b^- \;;\; s^+/s^-) \times \text{male } (b^+/Y \;;\; s^+/s^-)$$
$$(bald \text{ female} \times \text{wild-type male})$$

This will generate eight classes of offspring:

F1 females		F1 males	
genotype	**phenotype**	**genotype**	**phenotype**
$b^+/b^- \;;\; s^+/s^+$	wild–type female	$b^-/Y \;;\; s^+/s^+$	*bald* male
$b^+/b^- \;;\; s^+/s^-$	wild–type female	$b^-/Y \;;\; s^+/s^-$	*bald* male
$b^+/b^- \;;\; s^+/s^-$	wild–type female	$b^-/Y \;;\; s^+/s^-$	*bald* male
$b^+/b^- \;;\; s^-/s^-$	wild–type male	$b^-/Y \;;\; s^-/s^-$	*bald* male

Therefore the autosomal sex-determining gene confers femaleness, and when the gene is homozygous mutant (s^-/s^-) in a chromosomal (XX) female, the Titanians are sterile males.

Question 11: Without resorting to any molecular genetic analyses, how could you prove that Confused-1 and Confused-2 sterile mutant males have mutations in the same gene? (5%)

Answer 11: The standard route by which you would prove that two mutations are in the same gene would be to do a complementation test. Normally this involves crossing two mutants together and seeing whether the F1 offspring are mutant (in which case they are mutations in the same gene) or wild-type (in which case they are mutations in different genes). However, here the homozygous males are sterile. Nevertheless, you can cross fertile heterozygous Confused-1 males and heterozygous Confused-2 females together (and/or the reciprocal cross).

If they have mutations in different genes (gene A and B) then this will generate equal numbers of wild-type males and wild-type females, as every individual has at least one dominant wild-type allele at each gene.

		Heterozygous for A	
		$A^+ B^+$	$A^- B^+$
Heterozygous	$A^+ B^+$	$A^+A^+ B^+B^+$	$A^+A^- B^+B^+$
for B	$A^+ B^-$	$A^+A^+ B^+B^-$	$A^+A^- B^+B^-$

If they have mutations in the same gene (gene A) then there will be a ratio of 5:3 males to females, as one in four females is a homozygous mutant and will appear male.

		Heterozygous for A	
		A^+	A^-
Heterozygous for A	A^+	A^+A^+	A^+A^-
	A^-	A^+A^-	A^-A^-

Question 12: Speculate on how the gene is alternatively spliced in males and females. (5%)

Answer 12: There must be a sex-specific aspect to the slicing machinery. It's easiest to explain in females where a female-specific RNA-binding protein could bind to the 5′ end of exon 2, thus preventing its inclusion in the transcript. Alternatively there may be a male-specific RNA binding protein in males that makes the 5′ end of exon 2 appear to be (and therefore act like) a splice site. There are quite a few possible answers to this question, but any model has to accommodate the fact that mutant alleles in *confused-1* and *confused-2* are recessive mutations.

Question 13: According to the data presented in the Western blot in Fig. 3.5.10, what part of the Confused protein might the antibody targeted at? (5%)

Answer 13: The antibody has to recognise both the male and female versions of the protein. As the only thing they have in common is the 5′ end of the gene, the antibody must recognise the N-terminal of the protein.

Question 14: Speculate on how the *confused* gene might determine sex in Titanians? (10%)

Answer 14: You know that when a Titanian is homozygous for the recessive *confused* mutation it is male. Therefore the *confused* gene encodes a protein that confers femaleness. According to the data in Fig. 3.5.9, the *confused-2* mutation must be a null, producing absolutely no protein (functional or otherwise) as it produces no mRNA. Therefore, it looks like a simple loss-of-function mutation. The Western blot supports this as no Confused protein is produced in the *confused-2* mutant strain.

In males and females the gene is alternatively spliced. Female transcripts exclude exon 2, whereas male transcripts include exon 2. Yet you know that the shorter female-spliced mRNA makes a functional protein whereas the longer male-spliced mRNA does not. Therefore exon 2 must ensure that the male-spliced mRNA produces a non-functional protein. The most plausible explanation is that there is an in-frame stop codon in exon 2. Note that this is *not* a mutation. This is in the wild-type version of the gene in both males and females, but is effectively spliced out in females. This would explain the differences in female and male splicing patterns. The Western blot supports this hypothesis, as the male protein is significantly smaller than that in the female.

The *confused-1* splicing pattern in the sterile male is male-like, even though that male is chromosomally a female. This suggests the *confused-1* mutation affects the splice-site, so mutant females cannot splice-out (exclude) exon 2. This is supported by the Western blot data as the *confused-1* sterile males produce a protein of the same molecular weight as a wild-type male.

Question 15: If the *confused* gene encoded a transcription factor, how might it determine sex? How would you test this theory experimentally? (10%)

Answer 15: The Confused protein promotes femaleness; therefore, if it is a transcription factor, it might up-regulate female-specific genes or down-regulate male-specific genes. You could test this by microarrays or RNA-seq, comparing gene expression levels between wild-type and *confused* mutants. There are lots of possible answers here, but the key is that whatever experiments are proposed, you must compare wild-type and mutant Titanians.

Chapter 3.6: Once Bitten, Twice Shy

Question 1: What is the evidence supporting Dr Smith's hypothesis that 'mild late-onset zombieism' within the pedigree shown in Fig. 3.6.5 is an autosomal dominant inherited condition? Are there any problems with this hypothesis? (5%)

Answer 1: The Romero pedigree in Fig. 3.6.5 suggests that the condition is autosomal dominant. Every individual that is affected has one affected parent, and both males and females are affected. In addition, fathers (such as II:4) can pass the condition onto their sons (such as III:4), so it cannot be X-linked. The only problem with this theory is that male III:6 is unaffected yet he has an affected father and two affected sons, so the condition seems to 'skip' a generation.

Question 2: What, if any, is the significance of the fact that individual III:6 died at 29 years of age in an incident that involved three as yet unidentified individuals, a chain saw and an oven set at 210°C? (5%)

Answer 2: To be a dominant condition you would have expected male III:6 to be an affected heterozygote as his father and two of his sons are affected. However, as this is a late-onset condition and he has died in unfortunate circumstances at a fairly young age (29 years), he may have been too young to show any symptoms.

Question 3: How would you use the DNA sequence data from the 23 completely and accurately sequenced genomes, to identify the mutation that causes the mild-late-onset, zombie-like phenotype identified by Dr Smith? What problems might you encounter? (10%)

Answer 3: Assuming that the phenotype is caused by a dominant mutation, you are looking for a DNA change that is found in a heterozygous state in the ten affected members of the pedigree, but not in the 13 unaffected members of the family from whom you have DNA samples. Whilst this will exclude the vast majority of the genome and narrow it down to an area of one specific chromosome, this area may still be quite large, it may include many genes and, more importantly, it may include many DNA changes that are co-inherited with the causal DNA change. The key is determining which mutation is causal. Significant changes, such as nonsense mutations and frame-shift insertions/deletions would be good candidates. Missense mutations or non-coding mutations are harder to prove, as their consequences could be anything from catastrophic to silent. Other problems might include misdiagnoses, specifically false positives (someone apparently showing symptoms but actually unaffected or affected but by a different but phenotypically similar condition) or false negatives (pre-symptomatic, or mildly symptomatic individuals that have the mutation but no obvious mutant phenotype).

Question 4: If you could check one base pair of DNA sequence per second, how long in years would it take you to check the entire six billion (6 000 000 000) base pairs of a diploid human genome? (5%)

Answer 4: If you were to check one base pair of a 6 billion base-pair diploid genome at the rate of one base pair per second, the 6 billion seconds would precisely equal (6 000 000 000/60) 100 million minutes. This can be converted to (100 000 000/60) 1 666 667 hours, and similarly approximately (1 666 667/24) 69 444 days, which is just over (69 444/365.25) 190 years.

Question 5: What is the evidence supporting Dr Howe's assertion that the family shown in Fig. 3.6.7 has a rare, severe, autosomal recessive, early-onset, inherited form of zombieism? Why might this particular pedigree reveal an autosomal recessive genetic condition? (5%)

Answer 5: Autosomal recessive conditions suddenly appear in families, apparently without any warning or prior history, as any individual needs to be homozygous for a mutation to show a phenotype. As most mutations are very rare this is very unlikely to happen. However, couples related by descent (consanguineous couples) may well both be carriers (heterozygous) for the same recessive mutation, and would on average have one in four affected children. The fact that two siblings (V:2 and V:4) are affected here may be linked to the fact that their parents (IV:4 and IV:5) are cousins.

Question 6: What is the most plausible explanation for inheritance pattern of the T193I mutation in the pedigree identified by Dr Howe? (5%)

Answer 6: As the mutation is present in two of five children, but not in their parents, and as both affected children are heterozygous the mutation must be newly arisen and dominant. It presumably arose in the germ line of one of their parents.

Question 7: According to the data in Table 3.6.1, what is the most likely source of the apparent zombieism outbreak? (5%)

Answer 7: The only activity that all the students have in common is eating beef burgers from the refectory, so it seems likely that this must be the source of the problem. It therefore seems likely that the beef burgers are contaminated in some way.

Question 8: If none of the affected students have mutations in their *Zapped* gene, explain why they are apparently showing precisely the same symptoms as those individuals that do have mutations in their *Zapped* gene. (5%)

Answer 8: You have already established that the inherited form of zombieism described earlier is associated the *Zapped* gene, which encodes a prion protein. However, these students have no mutations in their *Zapped* gene and seem to have acquired zombieism by eating contaminated beef burgers from the refectory. A reasonable hypothesis is that the beef burgers are contaminated with rogue prion proteins. Infection via prions doesn't require any genetic susceptibility on behalf of the host, as wild-type individuals can be infected, and show progression of the death. Infection with prions also never changes the host DNA. However, pre-existing host mutations/variations can cause resistance to prion infection or affect the speed of progression. Infectious, or rogue, prions convert normal host prions into rogue forms. It is the rogue prion that has the apparent neurotoxic activity that seems to lead to zombie-like symptoms.

Question 9: Use a chi-squared test to determine whether V127/G127 heterozygotes are more likely to survive zombieism than G127/G127 homozygotes. What, if anything, does that tell you? (10%)

Answer 9: Undertaking the chi-squared test first. There are 2500 individuals of whom 378 are heterozygous (VG) and 2122 are homozygous (GG). In addition 414 died of a zombie-like condition and 2086 have not. Therefore, if you are equally likely to survive zombieism irrespective of the genotype (our null hypothesis) you can generate the expected number in each of the four classes:

$$
\begin{aligned}
\text{Hom/Alive} \quad &= ((\text{Freq hom}) \times (\text{Freq alive})) \times 2500 \\
&= (0.8488 \times 0.8344) \times 2500 \\
&= 1770.6 \\
\text{Hom/Dead} \quad &= ((\text{Freq hom}) \times (\text{Freq dead})) \times 2500 \\
&= (0.8488 \times 0.1656) \times 2500 \\
&= 351.4 \\
\text{Het/Alive} \quad &= ((\text{Freq het}) \times (\text{Freq alive})) \times 2500 \\
&= (0.1512 \times 0.8344) \times 2500 \\
&= 315.4 \\
\text{Het/Dead} \quad &= ((\text{Freq het}) \times (\text{Freq dead})) \times 2500 \\
&= (0.1512 \times 0.1656) \times 2500 \\
&= 62.6
\end{aligned}
$$

It might be easier to think of this in the following way. The observed individuals with each phenotype (from original text):

	Hom	Het	Total
alive	1721	365	2086
dead	401	13	414
total	2122	378	2500

To calculate 'expected frequencies' you start with the four 'total' scores which are simply the number observed, divided by the total observed (for example, 2086/2500 = 0.8344). The four genotype/phenotype frequencies are derived by multiplying frequencies of genotype and phenotype together (for example, 0.8344 × 0.8488 = 0.7082).

	Hom	Het	Total
alive	0.7082	0.1262	0.8344
dead	0.1406	0.0250	0.1656
total	0.8488	0.1512	1

To calculate the expected number of individuals, simply multiply the expected frequencies by 2500.

	Hom	Het	Total
alive	1770.6	315.4	2086
dead	351.4	62.6	414
total	2122	378	2500

The chi-squared calculation then takes the form:

Hom/Het Alive/Dead	Observed Number	Expected Number	O – E	(O – E)²	$\dfrac{(O-E)^2}{E}$
Hom/Alive	1721	1770.6	−49.6	2460.16	1.39
Hom/Dead	401	351.4	49.6	2460.16	7.00
Het/Alive	365	315.4	49.6	2460.16	7.80
Het/Dead	13	62.6	−49.6	2460.16	39.30
Total	2500	2500			55.49

Remember that for one degree of freedom the 5% significance limit is 3.84. Therefore, having a chi-squared value of 55.49 suggests that there is a highly significant difference in survival rates between genotypes. Heterozygous V127/G127 individuals are far less likely to die of zombie-like symptoms than homozygous G127/G127 individuals. It follows that the V127 allele confers some kind of partial, but significant resistance.

Question 10: Where on the road from Princess Port to Saint Dominica is each of the derived alleles (V127 and V129) likely to have arisen? (5%)

Answer 10: The most plausible explanation is that the alleles arose where they currently are at their most frequent. Therefore V127 probably arose in Princess Port, whilst V129 probably arose approximately 40 km west of Saint Dominica.

Question 11: How could you determine approximately when in history each of the derived alleles (V127 and V129), first appeared in the population? (5%)

Answer 11: As mutation rates and recombination rates are approximately constant, you can estimate the age of a mutation by studying the DNA sequence flanking the two derived alleles. When a new mutation (such as V127 or V129) first arises, every chromosome with that mutation will have precisely the same DNA sequence. However, over time recombination will strip away the flanking sequence, reducing the size of the sequences around the mutation that all mutant alleles share (that are identical in sequence by descent). Therefore, by studying the size of the flanking sequence that all mutant alleles have in common, you can estimate the age of the mutation.

Question 12: Give two plausible explanations why there do not seem to be any individuals that are homozygous for both derived alleles (V127 V129/V127 V129). (5%)

Answer 12: As the two derived alleles (V127 and V129) arose independently, the only routes by which they could become homozygous in the same individual are: (a) by recombination, even though they are just three base pairs apart; or (b) by a secondary mutation in one of the newly derived mutant strains. Both of these are extraordinarily unlikely events. Even if it did occur, it may be that the double mutation is homozygous lethal.

Question 13: Why might you be the only unaffected victim and why is Dr Howe so keen to take your blood sample back to the laboratory? (5%)

Answer 13: If you are the only ZITS employee that has been bitten but are unaffected, then presumably you carry a resistance mutation, possibly in the *Zapped* gene. You may have inherited this from your Great Uncle (or at least via his sister/brother who must be your grandmother/ grandfather) who was frequently exposed to the possibility of bites from zombies, yet survived to a grand old age. Clearly if you are resistant to zombieism, Dr Howe would be interested in the genetic basis of this resistance and in identifying any mutation (whether in the *Zapped* gene or another gene) that contributes to this.

Question 14: What is the difference between a treatment and a cure? (5%)

Answer 14: A treatment addresses the symptoms of a condition, and would usually need to be applied multiple times. A cure resolves the cause of a condition and (if carried out correctly) should only need to be applied once.

Question 15: Draw an annotated image of the construct you would use to allow you to make a mouse model of zombieism. Explain the function of each component of the construct, and discuss what problems you might you encounter and how might you resolve them. (10%)

Answer 15: The best model would replicate any of the human *Zapped* mutations precisely, and in this example you'll consider the D178N mutation.

It would be best to design a knock-in transgenic mouse that would result in it having the endogenous mouse *Zapped* gene (D178) with the equivalent of the N178 mutation. In addition to the N178 mutation, the construct (pictured here) would have a selectable marker, such as neomycin resistance (*neo^R*), within an intron adjacent to the target amino acid 178. It would also have a second selectable marker, such as thymidine kinase (*tk^+*), outside the region of interest.

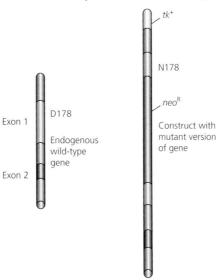

When this construct is introduced to D178 cells lines there are three possible outcomes: Targeted insertion: cells are *neo^R* and *tk^-*.

Random insertion: cells are *neo^R* and *tk^+*.
No insertion: cells are *neo^S* and *tk^-*.
For a knock-in clearly the targeted insertion is what you'd be selecting for.

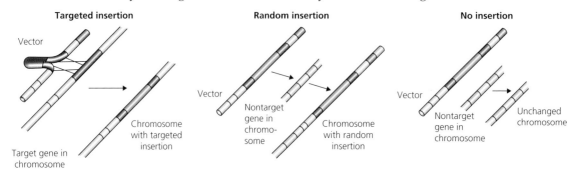

One problem here might be that the mouse and human *Zapped* genes have diverged over evolutionary time, so a D178N mutation either has no effect or may even be lethal. You could try different causative *Zapped* mutations, or even try to replace the mouse *Zapped* gene with the human *Zapped* gene.

In addition, the D178N mutation is late onset and mice have a much shorter lifespan than humans, so even if the mouse *Zapped* gene functions in much the same way as the human *Zapped* gene, it may not have time to develop the zombie-like phenotype.

Question 16: Having created a mouse model of zombieism, explain how a gene therapy approach might enable you to develop a treatment or cure for zombieism. What problems might you encounter and how might you resolve them? (10%)

Answer 16: As the D178N *Zapped* mutation is dominant, the objective would be to destroy the rogue prion gene and/or protein in the host. A key feature of the prion protein is its robustness, so it may be better to target the mRNA. In this case RNAi targeted at the mutant *Zapped* mRNA may remove this from the host cells. The problems here are:

(a) making sure the RNAi only flags mutant *Zapped* mRNA for degradation. In other words in heterozygous individuals the normal wild-type *Zapped* allele encodes the normal prion protein that presumably has an important function. So the *Zapped* RNAi construct would need to target the mutant allele, not the wild-type allele. Clearly this is less of problem if being homozygous for a null *Zapped* allele is not significantly deleterious. The key to this is the specific *Zapped* DNA sequence of the RNAi construct.

(b) making sure the *Zapped* RNAi only targets the *Zapped* RNA for degradation and not other mRNAs from genes with similar sequences (off-target degradation). Again, the key to this is the specific *Zapped* DNA sequence of the RNAi construct.

(c) getting RNAi expression at a high enough level to remove enough mRNA from the cell, and perhaps more significantly how the RNAi is going to be delivered to the target cells. You could partly address this by using a strong promoter, but getting the *Zapped* RNAi construct delivered to neural cells is always going to be a problem.

(d) ensuring RNAi expression occurs early enough to knock-down gene expression before the rogue Zapped protein reaches a critical level. This would require early and possibly pre-emptive (pre-symptomatic) expression of the *Zapped* RNAi.

There are lots of plausible alternative answers here.

Chapter 3.7: Red-Crested Dragons of Mythological Island

Question 1: Briefly describe the evolutionary relationship between the seven previously described dragon species shown in Fig. 3.7.4. (5%)

Answer 1: The eight dragon species are all derived from a single common ancestor. There seem to be three main evolutionary groups: a far east group (Japanese, Vietnamese and Chinese) that split from the Eurasian species, and two groups of Eurasian species that split more recently into western (Greek, Slavic and European) and eastern (Indian & Persian) groups. The red figures represent the probability that the node is correctly positioned. You can therefore be very confident in the position of high-scoring nodes (in the high 90s) but less confident about nodes that have lower scores.

Question 2: Redraw Fig. 3.7.4 to show the evolutionary relationship between the red-crested dragons of Mythological Island and the other eight dragon species. (5%)

Answer 2: The red-crested dragon shows the highest similarity scores with the three western populations (Greek, Slavic and European), so it probably belongs to that group. The best fit would look something like the modified Fig. 3.7.4 shown here.

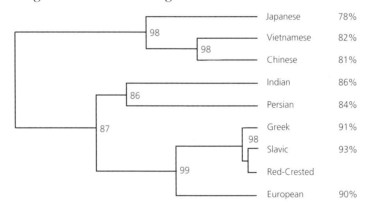

Question 3: If, as the evidence from Table 3.7.1 suggests, female dragons show a clear preference for males with red crests, speculate on why there are significant numbers of brown-crested and green-crested dragons in the Mythological Island population. How could you test your theory? (10%)

Answer 3: There must be an alternative, non-sexual, advantage to being a brown-crested or green-crested male. For example the brown and green colours may be useful camouflage, allowing the dragons to blend into their environment and avoid predation by the great mythological bustard, in a way that their red-crested brothers cannot. You could test this by directly observing predation of red-crested dragons by the great mythological bustard in the wild. Alternatively you might want to build models of red-crested dragons, place them in the forests of Mythological Island and measure the frequencies with which great mythological bustards attack the models. The second experiment is better for two key reasons: firstly, it eliminates any possibility of crest colour affecting dragon behaviour; secondly, it allows the experimenter to control for frequency of specific crest colours and other variables. Other possible explanations are possible, including models that develop the idea that variation at the crest colour gene is pleiotropic and has an as yet unknown second function that may even manifest itself in females (although you haven't yet explored the genetic basis of crest colour).

Question 4: Redraw Fig. 3.7.6 showing the crest colour genotype of each individual. Where genotypes cannot be absolutely determined, explain why. In addition, indicate one position within the pedigree where there is evidence of recombination between the crest colour gene and the STR. (10%)

Answer 4: Here is the redrawn Fig. 3.7.6:

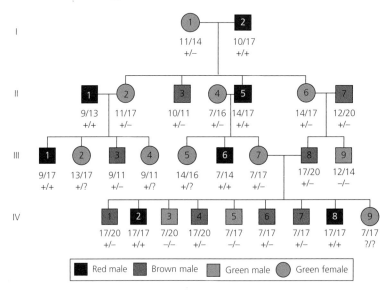

The crest colour genotypes of all males are known as the genotype reflects the phenotype (red is +/+, brown is +/−, green is −/−) where + is the red allele and − is the green allele. All females have green crests so their genotype can only be inferred by looking at their parents and off-spring. The four females here that have unknown genotypes (III:2, III:4, III:5 and III:9) have at least one heterozygous parent and no offspring.

Recombination is most clearly seen in the final generation. Parents III:7 and III:8 have three sons with the same STR genotype (7/17), two of whom are brown (+/−) and one of whom is green (−/−). At least one of them must be a recombinant.

Question 5: You now have the approximate map position of the crest colour gene. Describe in outline a strategy that will enable you identify the DNA sequence of the crest colour gene. (5%)

Answer 5: If you know the approximate map position of the crest colour gene you might com-pare DNA sequences in this region between red-crested, brown-crested and green-crested drag-ons. Whole-genome sequencing strategies will allow you to do this, but PCR will not as the region of interest (the length of DNA within which the mutation lies) is far too large. If you find a significant DNA change (such as a nonsense mutation, insertion or deletion or a frame-shift mutation) in this region, it would be sensible to study this further and, if the mutation is recessive, undertake gene rescue to prove it is the right gene. However, control region mutations are harder to identify, simply because there is usually more sequence variation outside the coding region. Similarly missense mutations where one amino acid is changed to another, may or may not simply cause the phenotype as the consequence of an amino acid change on protein function depends on the amino acids involved, their chemical properties, their position within the protein and its effect on protein function. It may be a good idea to narrow down the region of interest by mapping the mutation more accurately.

Question 6: Speculate on what kind of protein the crest colour gene may encode. Explain how it may generate the three distinct crest colour phenotypes in males, but only one crest colour phenotype in females. (5%)

Answer 6: The crest colour gene could encode an enzyme in a pigment pathway that converts a green pigment to a red pigment. In a red-crested (+/+) male both alleles make enzyme, and this is sufficient to convert a lot of, or maybe all of, the green pigment to red. In heterozygous (+/−) brown-crested males, one allele is functional and the other is mutant, so less enzyme is made and less green pigment is converted to red; the crest is a mixture of green and red and appears brown. In green-crested (−/−) individuals both alleles are mutant, so no functional enzyme is made and no green pigment is converted into red, so the crest remains green. As females only have green crests, you might expect the gene to only be switched on in males. So there may be a sequence in the control region of the gene encoding crest-colour enzyme that binds a male-specific transcription factor which switches the gene on in males, or a female-specific transcription factor which switches the gene off in females.

Other plausible answers are possible.

Question 7: Speculate on the nature of the mutations that causes green coloured crests in males from the north of the island (Fig. 3.7.7b) and the south of the island (Fig. 3.7.7c & Fig. 3.7.7d)? (10%)

Answer 7: Taking each set of RNA-sequencing data in turn:
  (a) The wild-type gene comprises five exons.
  (b) Green mutants from the north of the island are missing exon 2. The most plausible explanation is that there is a mutation at the beginning of exon 2 that results in it being excluded from the mRNA at splicing. Alternatively there may be a deletion of exon 2.
  (c) Green mutants with the south 1 version of the gene have a normal intron/exon structure and (at least on this limited data set) normal mRNA expression levels. Therefore the mutation could affect the quality of the protein (presumably via a missense mutation or a mutation affecting translational efficiency perhaps due to a deletion of the ribosome-binding site). It can't be a nonsense mutation in exons 1, 2, 3 or 4, but could be a nonsense mutation in exon 5.
  (d) Green mutants with the south 2 version of the gene have no detectable mRNA suggesting a control region mutation or nonsense-mediated decay due to a nonsense mutation in exons 1, 2, 3 or 4.

Question 8: What, if anything, does the RNA-sequencing data tell you about the origin of the green-crested dragons in the north and south of Mythological Island? (5%)

Answer 8: Three different green mutations on the island suggest three independent origins of the green phenotype.

Question 9: Why is Professor St George right to be sceptical of the theory that the decline of fire-breathing ability in red-crested dragons is due to the catastrophic failure of the Mythological chilli? (5%)

Answer 9: If the catastrophic decline of the Mythological chilli plant is the cause of the reduction in red-crested dragon fire-breathing activity, you would expect the decline in fire-breathing in red-crested dragon populations to mirror the geographic pattern of the failure of the Mythological chilli plants. However, the decline in the population of Mythological chilli plants is described as being catastrophic and originating in the west of the island, whereas the decline in red-crested dragon fire-breathing is gradual and originated in the south of the island. Therefore the geographical evidence strongly suggests the decline in Mythological chilli is not the cause of the decline in red-crested dragon fire-breathing ability.

Question 10: Propose a model for the genetic basis of the extinguished phenotype. (5%)

Answer 10: This could be a consequence of variation at a single gene (let's call it 'extinguished' E) where there are two alleles, one of which enables fire-breathing (E^+) and the other inhibits it (E^-). The normal control red-crested dragons from the north of the island would be E^+/E^+ and as a consequence have a high DAFT score (26–40). Extinguished (non-fire-breathing) dragons would be E^-/E^- and have an undetectable DAFT score (<5). Therefore the parents of extinguished dragons must be heterozygous (E^+/E^-) and as they have an intermediate DAFT score (11–25), the two alleles can be said to show intermediate dominance. It follows that the siblings of an extinguished dragon would segregate in a 1:2:1 ratio as shown. Therefore fire-breathing ability seems to be associated at a single gene that exhibits intermediate (incomplete) dominance.

Question 11: What do the data in Table 3.7.3 and Table 3.7.4 tell you about the *extinguished* mutation? How will you use this data to help you find the *extinguished* gene? (10%)

Answer 11: The *extinguished* (non-fire-breathing) allele (E^-) correlates almost perfectly with the green south 1 allele. In other words an E^-/E^- *extinguished* dragon is nearly always homozygous for the green south 1 allele, an E^+/E^- intermediate DAFT-activity dragon is nearly always heterozygous for the green south 1 allele, and wild-type high DAFT-activity dragons never have a green south 1 allele. Therefore, the extinguished mutation presumably arose in a dragon that already had the green south 2 mutation, and the two genes are very closely linked, as recombinants are very rare. Therefore you know the approximate map position of the *extinguished* gene and many of the arguments used in the answer to question 5 can be reiterated here.

Question 12: What are the normal expression patterns of the *extinguished* and *ill* genes? Speculate on why red-crested dragons may have evolved this pattern of expression. (5%)

Answer 12: In juveniles the *ill* gene is expressed at high levels but the *extinguished* gene is not. This is reversed in adults where the *extinguished* gene is highly expressed but the *ill* gene is not. The logic here may be that juveniles are protected against bacterial infection (*ill* on), but do not breathe fire (*extinguished* off) which would be dangerous to immature dragons, and for their friends and family. In adults *ill* is off so they have a compromised immune system, but as *extinguished* is on in adults they cook their food, which protects them from infection, and, at least in males, breathe fire to enhance their reproductive success.

Question 13: How are the expression patterns of the *extinguished* and *ill* genes altered in *extinguished* mutant juvenile and adult red-crested dragons? What consequences may these changes have for the phenotype of mutant red-crested dragons? (5%)

Answer 13: In the *extinguished* mutants, the expression patterns are reversed, so *extinguished* is expressed in juveniles and *ill* in adults. From an infection perspective this may not be a problem as fire-breathing juveniles can cook their food (though may not do this very efficiently) and adults have the *ill* gene expression. However, fire-breathing in juveniles may be dangerous and the lack of fire-breathing in adult males may inhibit their sexual activity. Therefore whilst *extinguished* mutants may have normal survival rates with respect to adult susceptibility to infection, juveniles may be more susceptible to infection (not cooking food properly) and accidents (associated with inappropriate fire-breathing), and adult males may not mate as frequently.

Question 14: Speculate on how a single mutation could cause the changes in *extinguished* and *ill* gene expression seen in the *extinguished* mutants. (5%)

Answer 14: This could be an inversion of the control region between *extinguished* and *ill*, with each acquiring the expression pattern of the other.

Question 15: Design a series experiments that would allow you to investigate further the continuing problem of population decline in red-crested dragons. (10%)

Answer 15: There are many experiments that could be undertaken:

Sequencing of the control region between *extinguished* and *ill* to determine the nature of the mutation.

Making transgenic red-crested dragons with: (a) reporter constructs to investigate *extinguished* and *ill* expression patterns in wild-type and *extinguished* mutants; (b) attempting gene rescue to prove these are the right genes.

Investigating the expression patterns and phenotypes of *extinguished* heterozygotes who you might expect to express both *extinguished* and *ill* all the time and therefore show heterozygous advantage.

Using qPCR to prove the microarray analysis is correct.

Using a knock-in transgene flanked by *loxP* site so that the gene is only expressed in adult or juvenile.

Using CRISPR to modify the *ill* or *extinguished* genes to test a named hypothesis

The key here is to make sure any answer reds like a proper experiment.

Chapter 3.8: I Scream!

Question 1: Redraw the Cornetto family tree to reveal the evidence supporting the notion that the Cornetto curse is a genetically inherited condition. In particular speculate on why individuals IV:5 and IV:8 show muscle weakness, muscle atrophy and abnormal muscle morphology. (5%)

Answer 1: The Cornetto curse (cold sensitivity) looks like an autosomal dominant condition, as it appears in every generation, affects males and females in similar frequencies (five females and six males) and can be passed from father to son (unlike X-linked conditions). Individuals IV:5 and IV:8 may be homozygous for the condition. If you called the dominant mutant allele + and the wild-type unaffected allele –, the genotypes of each family member would be:

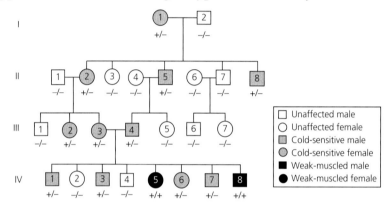

Question 2: Individuals III:3 and III:4 are married, but how else are they related and how does their additional relationship help you resolve the genetic basis of the Cornetto curse? (5%)

Answer 2: Cold-sensitive individuals III:3 and III:4 are cousins, as cold-sensitive II:2 (the mother of III:3) and cold-sensitive II:5 (the father of III:4) are siblings, both being the offspring of cold-sensitive grandmother I:1. This will help us find the mutant gene as all the dominant cold-sensitive mutations within the family are identical by descent. It follows that the two weak-muscled individuals (IV:5 and IV:8) are probably homozygous for identical mutant alleles inherited by descent. A key point is that two unrelated individuals are very unlikely to have identical mutations in the same gene, whereas two closely related individuals (such as cousins) may well do so.

Question 3: How can an understanding of the Cornetto family pedigree and the whole genome DNA sequence of each family member help you identify the mutation that causes the reversible cold-sensitive paralysis? What problems might you encounter? (5%)

Answer 3: Irrespective of whether the cold-sensitivity phenotype is caused by a mutation in a sodium channel gene or not, the strategy is essentially the same, although the detail of the investigation will be very different. It should be clear from the answer to question 1 that you know the genotypes of every member of the family, as there are three phenotypes: people that are unaffected (homozygous for the wild-type allele), people that are cold-sensitive (heterozygous for the mutant allele), and people that have profound muscle weakness (homozygous for the mutant allele). In addition, it seems that all cold-sensitivity mutations are derived from the cold-sensitive female I:1 and are therefore identical by descent, meaning all mutant cold-sensitive alleles in this family have exactly the same DNA mutation. Therefore you can simply compare the whole genome sequence between homozygous, heterozygous and unaffected individuals and look for mutations that are in both alleles of homozygotes, one allele of heterozygotes and neither allele of unaffected individuals.

Essentially you are looking for a DNA sequence change (mutation) that is absent in unaffected people (homozygous for the wild-type allele), in one dose in cold-sensitive people (heterozygous for the mutant allele) and both alleles in people that have muscle weakness (homozygous for the mutant allele). As there are 25 individuals in the family (and assuming you can get DNA samples from most of these), at the very least you should be able to narrow down the region where the mutation lies to a small region of one chromosome. Of course, all of this assumes that the family tree as drawn is correctly from a genetic perspective, though you have no evidence to suggest that it is not; but you should also check this. However, the region identified as being the source of the mutation causing the cold-sensitive phenotype may contain several mutations that have been co-inherited with the causal mutation. The problem will be identifying which of these mutations is causal. You could begin to address this by asking whether and how the mutation is predicted to affect (or even known to affect) protein function. You might also ask whether any of the correlating mutations are found in the wider population and if so whether they cause phenotypes in other individuals. This could be further complicated if the condition has varying penetrance and/or variable age of onset.

Question 4: How could you estimate the approximate date at which the mutation originally arose in the Cornetto family? (5%)

Answer 4: When a mutation first arises it is simply a DNA change in a specific region of a specific chromosome. Over time (generations), two things happen: mutation and recombination. In other words, the mutant chromosome acquires new mutations and also undergoes recombination, introducing new DNA sequence to the specific individual chromosome hosting the causal mutation. From a recombination perspective, the region around the causal mutation that came from the original chromosome will eventually become smaller and smaller, and in a sense the size of this 'common' region reflects the age of the mutation. Similarly, the number of 'new' mutations near to the causal mutation also reflects the time that has passed since the original causal mutation event. These regions of DNA are called 'haplotypes'. The age of the mutation can be estimated using your knowledge of the frequency of mutation and recombination, along with an understanding of the amount of DNA variation in the population (associated with its inbreeding/outbreeding status).

Question 5: What is the consequence of each of the five DNA sequence changes in the cold-sensitive allele, and which may be responsible for the cold-sensitive phenotype in the Cornetto family. (5%)

Answer 5: Taking each of the five DNA sequence changes in turn:

Codon 1273	CTC to ATC	Leucine to Isoleucine (L1273I)
Codon 1278	TTC to TTT	Phenylalanine (no change)
Codon 1288	AAC to AAT	Asparagine (no change)
Codon 1305	TTA to CTA	Leucine (no change)
Codon 1306	GGG to AGG	Glycine to Arginine (G1306R)

There are three silent mutations that do not alter the amino acid in the protein, so they are unlikely to affect protein function.

Either of the two missense mutations, L1273I and G1306R, could be the cause of the phenotype.

Question 6: Why does the *P*-element transgene contain: (a) the 31 bp inverted repeats; and (b) the wild-type copy of the *white* eye colour gene? (5%)

Answer 6: Taking each component in turn:

The 31 bp repeats facilitate the insertion of the *P*-element (including the transgene) into the host chromosome. The 31 bp repeats define the ends of the *P*-element, and the ends recombining with the host DNA enables the *P*-element (and therefore all the sequence between the 31 bp repeats) to insert onto the host's DNA.

The wild-type copy of the *white* eye colour gene is dominant and confers a red eye colour on the host fly (remember that *Drosophila* genes are named after their mutant phenotype), therefore acting as a marker for the presence (red eye) or absence (white eye) of the transgene in the potential host fly. Clearly any potential host must be a *white* eye mutant.

Question 7: What transgenic flies are you going to make to identify the DNA sequence change (mutation) in the human SCN4A gene that causes the Cornetto cold-sensitivity phenotype? Specifically, what DNA sequence(s) should the *Gene of Interest* in the *P*-element (see figure 9) actually have? What problems might you encounter and how might you resolve these? (5%)

Answer 7: The Cornetto cold sensitivity seems to be caused by one (or both) of two missense mutations in human SCN4A, namely L1273I and G1306R. It would therefore be sensible to make four transgenes, one containing both missense mutations I1273 R1306 (which you would predict to be cold-sensitive, so is a positive control), each mutation on its own, specifically I1273 G1306 (which may or may not be cold-sensitive) and L1273 R1306 (which may or may not be cold-sensitive) and a wild-type control L1273 G1306 which should not be cold-sensitive. As you don't know whether the human SCN4A gene is functional in *Drosophila*, you should mutate the equivalent bases in the fly gene. As there is a limit to the size of transgenes, you will have to modify a spliced *paralytic* gene, so the transgene will not be subject to alternative splicing and may not undergo RNA editing (depending on where the RNA editing signals are). There is no simple way of resolving this, but it would be sensible to use a common *paralytic* splice variant to modify for the transgene. The transgene should be expressed under its own (*Drosophila paralytic*) control region.

Question 8: When you have made your *P*-element transgene you will need to inject it into the germ line of a host fly. That host fly could be wild-type, heterozygous or homozygous mutant for the *paralytic* gene, but what would be the best *paralytic* genotype and why? (5%)

Answer 8: The host fly must be homozygous for the wild-type *paralytic* allele. There are two reasons why the host cannot be a *paralytic* mutant. Firstly, null *paralytic* mutants are recessive lethal, so flies homozygous for the *paralytic* mutation (*para/para*) die. In addition, the cold-sensitive SCN4A mutation in the Cornetto family is dominant so you might expect to see (or at least hope to see) a cold-sensitive phenotype in transgenic fly even if that host is homozygous wild-type at the *paralytic* gene. In any case, if you wanted to do so, once a transgenic line is established you can use crossing schemes to put the transgene in a *paralytic* mutant background.

Question 9: What does the cold-exposure assay reveal about the genetic basis of cold sensitivity? (5%)

Answer 9: The cold sensitivity of flies carrying the I1273 transgenes are indistinguishable from wild-type, in that none of them show any evidence of cold sensitivity. However, three of the four strains carrying the R1306 transgene are cold-sensitive. So it would seem that G1306R missense mutation is the source of the cold sensitivity.

Question 10: Even though transgenic lines R1306 (1) to R1306 (4) have a single copy of the same transgene, they show marked differences in their sensitivity to cold. Speculate on why this might be and design an experiment that would test your theory. (10%)

Answer 10: The four R1306 strains have identical genotypes and only differ in the insertion site of the transgene. Evidence from the Cornetto family tells us that R1306 is dominant, so it may be that the degree of cold sensitivity in these lines is associated with the expression levels of R1306 (this could be absolutely due to the level of expression of R1306, or the relative levels of expression of R1306 against the host transgene, although the level of expression of the host transgene in each of the fly lines should be similar). Different levels of expression of R1306 could be due to the different insertion sites of the transgene that presumably affect levels of expression of the R1306 *paralytic* transgenes. You could test this experimentally by measuring the levels of the R1306 *paralytic* transgene mRNA, perhaps by qPCR. The key issue here will be distinguishing between the R1306 *paralytic* mRNA and endogenous *paralytic* mRNAs. The qPCR approach would work if the final base in one of the primers sat on the missense base of R1306, such that only the transgene's mRNA is amplified. Similarly RNA-seq will work because you will know the relative frequency of the two varieties of alleles (transgene R1306 and endogenous gene G1306).

Question 11: How is an evolutionary tree such as that presented in Fig. 3.8.11 constructed? How can you be confident that its shape is reliable and robust? (5%)

Answer 11: Evolutionary trees are constructed by looking for sequence similarity at the DNA base (or perhaps amino acid) level. Sequences that show high similarities are more closely related. In Fig. 3.8.11 the lengths of the horizontal lines reflect the amount of similarity between sequences (so you can infer that SCN4A and SCN5A are the most similar sodium channels). Various statistical analyses can be employed to determine whether the shape of the tree is reliable and robust.

Question 12: Like humans, mice also have nine sodium channel genes. How could you determine which, if any, of the mouse genes is equivalent to the human SCN4A gene? What problems might you encounter? (5%)

Answer 12: The relationship between the nine human sodium channel genes and the nine mouse sodium channel genes will primarily reflect the status of those genes in their last common ancestor. So if the last common human/mouse ancestor had nine sodium channel genes, and each of these survives in extant human and mouse populations, then the nine human and nine mouse genes are absolutely equivalent and map one on one. Therefore, if you were to draw an evolutionary tree from all 18 genes, their sequences would reflect their evolutionary history and one mouse gene would be similar to human SCN1A, one would be similar to SCN2A, and so on. However, it is entirely possible that the last common ancestor has a different sodium channel gene number or that within the human and/or mouse lines, sodium channel genes may have been deleted or duplicated and will certainly have diverged.

Question 13: Design a construct that would allow you to make mouse with a 'knock-in' SCN4A R1306 mutation. Briefly describe the strategy that will allow you to generate a transgenic mouse with an SCN4A R1306 mutation. (10%)

Answer 13: The construct (targeting vector) could have the following structure:

In the upper image three introns (yellow) and two exons (blue) are shown. The G1306 codon is shown as being in Exon Q. The objective is to use homologous recombination to replace this exon with an Exon Q carrying R1306. To achieve this, the targeting vector has been constructed. It has part of Intron O-P (yellow), all of Exon P (blue), all of Intron P-Q (yellow) interrupted by a neomycin-resistance gene (green) that is flanked by loxP repeats (black), a modified R1306 Exon Q (blue), a short section of Intron Q-R, and a thymidine kinase wild-type gene.

You would introduce the targeting vector into wild-type (SCN4A G1306) mouse embryonic stem cells. There are three possible outcomes:

Homologous recombination – this is the desired outcome. Recombination occurs such that the endogenous chromosome takes up *neo*^R and R1306 Exon Q, but not *tk*^+. In cell culture these cells are resistant to neomycin but sensitive to ganciclovir.

No recombination – an undesirable outcome. There is no change to the host cell, and it remains neomycin and ganciclovir sensitive.

Random insertion – an undesirable outcome. In this case the entire vector is inserted so the cells are neomycin and ganciclovir resistant.

Cells that are neomycin resistant and ganciclovir sensitive should have a modified (R1306) SCN4A gene. You can implant these cells into pseudopregnant females and the resulting embryos should be SCN4A heterozygotes (G1306/R1306). The neo^R gene in the intron may or may not affect the expression of SCN4A R1306. If it does, then crossing SCN4A R1306 mice to mice carrying the CRE recombinase should create embryos in which the neo^R gene excises.

Note that other selectable markers are available.

Note that this is a stylized image and in the real human SCN4A the introns are not labelled P and Q. Other similar strategies that could achieve the same outcome are available.

Question 14: Speculate on why the cold sensitivity of the four original transgenic strains R1306 (1–4) has in some cases changed over the 18 months since they were last tested. (10%)

Answer 14: The four transgenic strains R1306 (1-4) have been kept in a homozygous state both for the endogenous SCN4A G1306 allele and for the SCN4A R1306 transgene. You know that originally three of these four strains were cold sensitive (though with different severities) and that over 18 months (approximately 36 generations) the cold sensitivity has largely disappeared. In strains with mutant phenotypes, selection may occur to reduce the severity of the mutant phenotype. (A good example of this is *Drosophila* strains homozygous for the *eyeless* mutation). Here, the easiest way to achieve this might be for selection to reduce the expression of the R1306 transgene relative to the endogenous G1306 gene. In the original transgenic line there will be small but significant differences in the relative expression of R1306 and G1306. It seems reasonable to suppose that flies with higher expression of G1306 relative to R1306 will be 'fitter' and produce more offspring. This capacity will be passed to the next generation, and over 36 generations may lead to a more-or-less complete loss of the cold-sensitive phenotype.

Question 15: Design an experiment that would allow you to identify whether the loss of cold sensitivity in the transgenic lines has a genetic or environmental cause. (5%)

Answer 15: If the lack of cold sensitivity is genetic then it must be inherited, so you could simply look at the cold sensitivity in subsequent generations. The level (or absence) of cold sensitivity should be consistent from generation to generation. This should be true within each of the four transgenic lines. Effectively the transgenic strains have been subjected to a selection experiment, so each strain may have accumulated many small genetic (allelic) variants that contribute to the loss of the cold sensitivity phenotype. If you outbreed each strain then these effects should disappear in the F1 and be diluted out in the F2.

Question 16: Design an experiment that would allow you to identify the genetic cause of the loss of cold sensitivity in the transgenic lines. (10%)

Answer 16: You are looking for differences between the original transgenic strains and the same strains 18 months (about 36 generations) later. The experiments described earlier reveal that there no DNA sequence changes (mutations) that affect the quality of one or more proteins produced. So the loss in cold sensitivity is more likely to be caused by changes in expression levels of specific genes. You could look at changes in expression levels over the 18-month period by undertaking a whole genome microarray analysis or undertake RNA-seq analysis on the transgenic R1306 (1) strain when originally constructed (assuming this RNA is in the freezer) and the transgenic R1306 (1)* strain existing today. In fact there are a number of ways that this could be investigated.

Acknowledgements

With thanks to Helen Lindsay.

Chapter 3.9: The Nuns of Gaborone

Question 1: Define the term 'species' and, in particular, explain how analyses of DNA sequences can be used to determine whether two populations should be regarded as one species or two. (5%)

Answer 1: A species can be defined as a group of individuals that can mate to generate viable and fertile offspring. From a genetic perspective you can compare populations at the level of DNA sequence variation. It may be that from a DNA sequence perspective two population groups are genetically distinct, so might be considered different species or sub-species. In reality determining what is, or is not, a species is perhaps harder than may at first appear.

Question 2: Is the decline in blue wildebeest populations primarily associated with the drop in rainfall, or is there likely to be another explanation? Explain. (5%)

Answer 2: As already stated, the statistical analysis suggests there is no correlation between population decline and rainfall. If the decline in blue wildebeest populations were primarily associated with decreased rainfall you would expect all populations to be affected in a similar fashion, but Fig. 3.9.4 clearly shows that is not the case as the Moremi, Kgalagadi and Mashatu populations seem to be less severely affected, yet the decline in rainfall in these populations is comparable to the other very severely affected populations. Therefore there must be another explanation.

Question 3: Explain why a virus that is endemic in a cattle population where it causes few health issues, might be potentially lethal when it crosses the species barrier and infects blue wildebeest? (5%)

Answer 3: When a new infective agent such as a virus infects a host organism that has never encountered that virus in its evolutionary history, then the new host is unlikely to have any natural immunity to it and it may have a devastating effect on the health of the new host. The new host will no doubt have some genetic variability in its response to the new virus, and only those individuals who by chance have some level of natural genetic resistance will survive and pass the survivor alleles to the next generation. Over a short time all animals in that population will be selected for the survivor alleles and all individuals will exhibit resistance or tolerance of the new virus, so whenever these animals, or their descendants, are exposed to the virus, none of them will be seriously affected by it. If the virus is then passed to a new species/population, the whole process would start over again.

Question 4: How could you determine that an infective agent such as a virus is linked to the recent decline in blue wildebeest population size? What problems might you encounter? (5%)

Answer 4: If you know which tissue is affected (and in this case it's implied that it is linked to immune system dysfunction) you could try to infect a previously uninfected animal with the contaminated tissue. In this case you might take blood from a sick blue wildebeest, purify the virus and inoculate a fit blue wildebeest and see whether it develops the symptoms. A key issue is how do you know that the fit wildebeest you are inoculating is not already infected but as yet asymptomatic? In the absence of a molecular test (see question 5) the only sensible way to do this would be to infect blue wildebeest that are perhaps in your research facility and have had no contact with, and have been isolated from, wild blue wildebeest and domesticated cattle.

Question 5: How could you determine whether blue wildebeest can be infected asymptomatically? (5%)

Answer 5: Once you had identified the specific virus-type involved (and in this case you might at first look at the virus that devastated cattle populations in the 1920s), you could develop a PCR-test that amplifies conserved DNA sequences from the virus within blue wildebeest samples.

Question 6: How could you determine whether the blue wildebeest were infected with the same virus that can infect cattle? (5%)

Answer 6: If the virus infecting blue wildebeest very recently came from cattle, then the virus in the blue wildebeest should show very high DNA sequence similarity with viral DNA samples taken from domesticated cattle.

Question 7: What do the data based on crosses between blue wildebeest within populations suggest about the genetic basis of resistance within the three populations? (5%)

Answer 7: These data are best explained by there being a single gene within a population determining resistance/susceptibility. In each case the susceptibility allele is dominant and the resistant allele is recessive.

Question 8: What do the data based on crosses between blue wildebeest of the different populations suggest about the genetic basis of resistance in Botswana blue wildebeest? (5%)

Answer 8: The genetic basis of resistance/susceptibility in the three populations is associated with the same gene.

Question 9: What does the RNA-seq data reveal about the genetic basis of resistance in the Moremi, Kgalagadi and Mashatu populations of blue wildebeest. (10%)

Answer 9: In the Moremi population there is a small deletion in the second exon. Deletions within coding regions often result in recessive loss-of-function mutations, so the deletion is the probable cause of resistance in the Moremi population. However, the RNA-seq analysis reveals no apparent difference in intron/exon structure the wild-type (sensitive) and Kgalagadi and Mashatu (resistant) blue wildebeest. Nevertheless, as you already know that the recessive resistance mutations in the three populations are in the same gene (as they fail to compliment), you know there must be a mutation within the Kgalagadi and Mashatu genes. However, this mutation does not affect RNA levels, so cannot be a control region promoter/enhancer mutation, nor is it likely to be a premature stop codon as you would expect non-sense-mediated decay (NMD) to eliminate, or at least reduce, RNA levels. In addition, the splicing pattern is normal so it is not a splicing defect. Therefore the possibilities are a missense mutation such that a non-functioning or reduced functioning protein is made, or a mutation that affects translation such as a defect in the translational start signal so no protein is produced.

Question 10: What does the DNA sequence data suggest might be the genetic basis of viral resistance in the Moremi, Kgalagadi and Mashatu populations of blue wildebeest? How would you test your theory and what alternative explanations might there be? (10%)

Answer 10: You already know that resistance is caused by mutations in the chemokine receptor gene. Taking each population in turn:

In the Moremi population, the RNA-seq data resistance is associated with a small deletion in exon 2. If you look at the modified table of codon differences given here and compare susceptible and resistant blue wildebeest in Moremi, you see two silent mutations (codons 7 and 222) and one missense mutation (codon 73). The codon 73 change from leucine to isoleucine almost certainly has no consequence as they are very similar amino acids. Apart from the small deletion found earlier, the susceptible and resistant Moremi wildebeest have identical sequences, which suggests that the deletion is responsible for the resistance.

In the Kgalagadi population, when you compare susceptible and resistant blue wildebeest there are two missense mutations: codon 33 (tyrosine to cysteine) and codon 336 (serine to threonine). One or both of these could cause loss of chemokine receptor function, but as tyrosine and cysteine have very different properties, whereas serine and threonine are quite similar, the codon 33 change is more likely to be causal.

In the Mashatu population, when you compare susceptible and resistant blue wildebeest there is a single missense mutation at codon 147 (cysteine to arginine). From an amino acid property perspective this is a big change, so is probably the cause of the loss of function of the chemokine receptor.

Codon	Moremi		Kgalagadi		Mashatu	
	susceptible	resistant	susceptible	resistant	susceptible	resistant
7	CCT	CCT	CCC	CCC	CCC	CCC
	Proline	Proline	Proline	Proline	Proline	Proline
33	TAC	TAC	TAC	TGC	TAC	TAC
	Tyrosine	Tyrosine	Tyrosine	Cysteine	Tyrosine	Tyrosine
34	CCC	CCC	CCC	CCC	CCA	CCA
	Proline	Proline	Proline	Proline	Proline	Proline
73	CTA	ATA	CTA	CTA	TTA	TTA
	Leucine	Isoleucine	Leucine	Leucine	Leucine	Leucine
134	AGT	AGT	AGC	AGC	AGT	AGT
	Serine	Serine	Serine	Serine	Serine	Serine
147	TGT	TGT	TGT	TGT	TGT	CGT
	Cysteine	Cysteine	Cysteine	Cysteine	Cysteine	Arginine
201	ATA	ATA	TTA	TTA	ATA	ATA
	Isoleucine	Isoleucine	Leucine	Leucine	Isoleucine	Isoleucine
222	CAT	CAT	CAC	CAC	CAC	CAC
	Histidine	Histidine	Histidine	Histidine	Histidine	Histidine
334	AGG	AGG	AGG	AGG	CGG	CGG
	Arginine	Arginine	Arginine	Arginine	Arginine	Arginine
336	TCG	TCG	TCT	ACT	TCT	TCT
	Serine	Serine	Serine	Threonine	Serine	Serine

The table shows the DNA sequence differences at 10 named codons in the chemokine receptor gene in susceptible and resistant blue wildebeest from the three populations (Moremi, Kgalagadi and Mashatu) and additionally shows the amino acids encoded by each codon.

Associations between SNPs and resistance provides an indication that the SNPs are causative, but they do not prove it. Functional experiments would be required to show that the T cells from animals with a certain genotype were susceptible or resistant to infection. It is possible, though in this case fairly unlikely, that the missense mutations are not causal but are co-inherited with the true mutation that leads to resistance.

Question 11: What does the DNA sequence data suggest about the evolutionary origin of resistance in Botswana's blue wildebeest? (5%)

Answer 11: The three populations have different resistance mutations, so these must be independently derived. At least in this very limited dataset, the three populations have very similar chemokine receptor DNA sequences. Indeed the number of DNA changes between the susceptible alleles is 5 (Moremi vs Kgalagadi), 6 (Moremi vs Mashatu) and 5 (Kgalagadi vs Mashatu). DNA differences between susceptible and resistant alleles within a population are much smaller: 4 (Moremi, including the deletion), 2 (Kgalagadi) and 1 (Mashatu). This supports the idea that the resistance mutations arose recently, and since the populations split.

Question 12: Suggest two possible mechanisms by which a blue wildebeest heterozygous for a chemokine receptor mutation may have partial resistance to viral infection. (5%)

Answer 12: Individuals that are heterozygous will be sensitive to the virus as they have functional receptors encoded by the wild-type allele, and the mutation is therefore recessive. However, it may be that heterozygotes exhibit partial resistance as they may have fewer receptors on the surface of their T cells than homozygous wild-type individuals because they only have receptors encoded by a single allele. This is effectively a dosage effect. Alternatively, the protein made by the mutant allele may interfere with proteins made by the wild-type allele. If, for example, the functional receptor protein is a homodimer, then a mutant non-functional protein may dimerize with a protein from the functional allele, making a non-functional dimer. This may mean that in a heterozygote not only is one allele making a non-functional receptor, but it is also decreasing the amount of functional receptor produced by the wild-type allele.

Question 13: Calculate the mean ligand gene copy number in infected and uninfected individuals. Do these data suggest there is any effect of ligand gene copy number on blue wildebeest survival rates when they are exposed to the virus? (5%)

Answer 13: The mean ligand gene copy numbers are 2.21 for infected blue wildebeest and 3.05 for uninfected blue wildebeest. If this is statistically significant it tells us that the fewer ligand genes there are in an individual, the more likely they are to be infected.

Question 14: For your application to WOBBLE, propose a model showing how variation at the chemokine receptor gene and chemokine ligand gene affect survival rates when blue wildebeest are exposed to the virus. (10%)

Answer 14: The virus infects the host via the chemokine receptor. If the receptor is non-functional because of a mutation in the chemokine receptor gene, as it is in some individuals within Moremi, Kgalagadi and Mashatu populations, then individuals that are homozygous for the receptor mutation will be resistant to the virus because it has no route by which it can enter the cell. There are many ways to test this. Western blots work well for membrane protein, though a very hydrophobic membrane-bound protein can cause problems. The key is having a good antibody. Alternatively you could use flow cytometry to compare relative amounts of receptor on the surface of T cells in wild-type, heterozygote and mutant individuals.

Variation in gene copy number in the chemokine ligand gene probably affects and correlates with affect the amount of ligand protein produced. To calculate the amount of ligand, the conventional high-throughput quantitative assay is an enzyme-linked immunosorbent assay (ELISA) that uses antibodies and colour change to identify the quantity of a specific protein. As the ligand binds to the chemokine receptor and reduces the amount of chemokine receptor on the surface of cells, it effectively prevents the virus from gaining access to the T cells. If more ligand protein is produced, then fewer functional receptors remain, and the rate of infection is reduced.

Question 15: For your application to WOBBLE, propose in outline how you might develop one treatment for infected blue wildebeest, and one preventative measure that may help blue wildebeest survive the viral infection. (5%)

Answer 15: Taking each question in turn:

A treatment targets the symptoms without addressing the underlying cause. In this case you could develop a drug that blocks the chemokine receptor and outcompetes the virus. This may slow down infection, but it probably won't prevent it. This approach could also be used prophylactically. So dose uninfected wildebeests with drugs that inhibit the chemokine receptor, and this should make them resistant to infection. You might want to mention the possibility of side effects.

A cure is a preventative measure that will need to address the cause of the problem and prevent infection getting a hold. This may be germ-line gene therapy that knocks out the receptor gene.

Question 16: What do the data indicate regarding perpetrator of the crime? (5%)

Answer 16: The STRs of the blood sample are a perfect match to Sister Anna. This places her at the crime scene, which may require some explaining, but does not necessarily mean that she committed the crime.

Question 17: In addition, several of the other people living in the hospice are related. Who are the twins, who are the siblings and who are the parent and child? And what are you going to tell your father? (5%)

Answer 17: Taking each pair in turn:

The twins are Sister Boitumelo and Sister Kagiso who have identical STR profiles and are the same age.

The siblings are Spyros and Maria who are therefore brother and sister and are similar plausible ages for siblings (25 and 23). They share precisely half their STRs.

Sister Andrea is your mother. You share half your STRs with Sister Andrea, precisely one at each STR, so she is extremely unlikely to be your sister.

Quite what you tell your father is up to you.

Acknowledgments

With thanks to Professor Rob Nibbs and Paula Blair for comments on this chapter.

Chapter 3.10: Poissons Sans Yeux

Question 1: Design a non-invasive experiment that would allow you to determine whether the psychofish populations in each tank live in a 24-hour world. (5%)

Answer 1: You already have information suggesting that if the fish can perceive the time of day then they might be more active during the day and less active during the night. You are told the fish are valuable, so you're not allowed to undertake any invasive molecular-type experiments that would kill them. In theory you could simply look at the fish and measure their activity, but that is impractical given the scale of experiment you need to undertake. The ideal experiment would involve developing an automated system to measure activity within the tanks of fish. This might involve placing lasers across the tank such that every time a beam is broken a detector scores that event. It follows that active fish will break the beam more frequently than less active fish. Assuming this can be recorded accurately, the frequency of broken beams per hour over a 24-hour period will reflect activity over the same 24-hour period. Any answer that allows scoring of activity automatically, such as using a camera linked to software that somehow scores movement in a non-invasive fashion, is good.

Question 2: What is the relationship, if any, between temperature and the activity rhythm in psychofish? (5%)

Answer 2: There is an inverse relationship between temperature and length of activity rhythm. In other words, the higher the temperature, the shorter the rhythm. Given that all other environmental issues are controlled for, it would seem that temperature directly influences the length of the activity rhythm. At 10°C the activity rhythm is about 24 hours, at 20°C it is about 22 hours (which matches the data in Fig. 3.10.4, which would presumably have been completed at room temperature of about 21°C) and at 30°C the activity rhythm is approximately 18 hours.

Question 3: How would you prove that the abnormal activity rhythm phenotypes found in psychofish from tanks 13, 17, 68 and 70 are genetic in origin? (5%)

Answer 3: As every psychofish within tanks 13, 17, 68 and 70 has the same abnormal phenotype strongly suggests that the abnormal activity rhythm phenotypes are genetic in origin. Nevertheless, to prove this is a genetic phenotype you need to demonstrate that the characteristics are inherited, rather than a property of some anomaly associated with the specific tanks. So you could swap the fish and tanks around and see whether the abnormal rhythm follows the fish, rather than the tank. This will require you to cross fish within a strain and show that the F1 (and subsequent generations) have the same mutant phenotype. You can investigate this further by crossing the psychofish from each of the four 'mutant' tanks (13, 17, 68 and 70) to other psychofish strains to determine whether the phenotype is inherited as a single gene condition (meaning in a Mendelian fashion). For example, if the mutant phenotype is autosomal recessive then crossing a mutant psychofish to a wild-type psychofish would generate an F1 of heterozygous psychofish with a wild-type phenotype. Crossing the F1 psychofish together would generate an F2 of three wild-type to one rhythm mutant.

Question 4: How would you use existing data from the Psychofish Genome Project, plus your new whole genome sequence data from the 100 tanks to identify the gene(s) associated with the mutant phenotypes shown in the four strains? What problems might you encounter, and how might you seek to resolve these? (5%)

Answer 4: As each tank was set up from a single pregnant female, you can reasonably conclude that psychofish within each tank are highly inbred and have relatively little DNA variation. Therefore if you compare the entire DNA genome sequences of strains 13, 17, 68 and 70 with each other and the sequences of the other 96 psychofish strains, you might be lucky and find a DNA variant that is unique to each strain. However, it's likely that there are several DNA variants unique to each of the four strains, so identifying the precise causal DNA change may be more difficult than would first seem. However, you might be lucky as the mutations in the four strains could all be in the same gene, and at least in one (or more) of the strains, the mutation could be fairly obvious, such as a stop codon, a deletion, an inversion or a translocation.

Question 5: Speculate on how the different types of mutation listed in Table 3.10.1 might cause the specific rhythm phenotypes shown in Fig. 3.10.5. (5%)

Answer 5: Taking each of the mutations in turn (and remembering that in earlier studies higher temperatures resulted in shorter rhythm, suggesting the higher activity of Greenwich protein causes shorter rhythm):

Strain 13 has a missense mutation that results in a short 18-hour rhythm. Therefore this amino acid change may cause the Greenwich protein to be more active or more stable.

Strain 17 has a nonsense mutation that results in an arrhythmic phenotype. Therefore the truncated Greenwich protein presumably has no function. Alternatively, if the nonsense mutation is in anything but the last exon, the mRNA may undergo nonsense-mediated decay (NMD) meaning no protein is produced.

Strain 68 has a missense mutation that results in a long 28-hour rhythm. Therefore this amino acid change may cause the Greenwich protein to be less active or less stable, but it cannot be non-functional.

Strain 70 has an insertion in an intron that results in an arrhythmic phenotype. This would suggest that the Greenwich protein is non-functional so the insertion probably affects splicing, either interrupting a key splice site signal, or introducing a new splice site signal.

Question 6: What should each of the three transgenes (T1, T2 and T3) reveal about the control of *greenwich* gene expression? In addition, when introduced to wild-type psychofish, why do each of the three transgenes show the circadian GFP expression pattern revealed in Fig. 3.10.7? (10%)

Answer 6: To take each transgene is turn:

Transgene 1 has the GFP coding region fused onto the control region of *greenwich*, so the expression of GFP should accurately reflect the expression of the endogenous *greenwich* control region. As GFP is expressed in a 24-hour pattern, these data strongly suggest that the *greenwich* gene is also expressed in a 24-hour rhythmic pattern. This kind of construct is often described as a transcriptional fusion.

Transgene 2 has the GFP coding region fused in-frame to a significant proportion of the *greenwich* coding region and driven by the *greenwich* control region. As with transgene 1, you would expect GFP to reflect the expression of the endogenous *greenwich* gene. Again, as GFP is expressed in a 24-hour pattern, these data strongly suggest that the *greenwich* protein is also expressed in a 24-hour rhythmic pattern. In addition, depending on whether any of the Greenwich protein included in the hybrid Greenwich–GFP protein retains any function, the GFP may be targeted to a specific region of the cell. This kind of construct is often described as a translational fusion.

Transgene 3 has a constitutively expressed control region fused to the coding region of GFP. This is simply a control that will show whether or not genes in psychofish are normally expressed in a rhythmic fashion. There is no evidence that this construct is expressed in a rhythmic fashion.

Overall you can conclude that *greenwich* mRNA and Greenwich protein are expressed in a 24-hour rhythmic pattern, and that this is determined by the control region of the *greenwich* gene.

Question 7: What is the significance of using a homozygous *greenwich* null mutant psychofish strain as a host for the transgenes? (5%)

Answer 7: As the host strain has null mutation, the host psychofish will produce no functional Greenwich protein. Therefore these data will compare transgene expression in the presence (wild-type) and absence (null mutant) of endogenous rhythmic Greenwich protein. Therefore this will tell you whether the rhythmic expression of the *greenwich* gene is in part associated with its own expression. In other words, it's testing for a feedback loop.

Question 8: What do the data in Fig. 3.10.8 tell you about the control of *greenwich* gene expression in psychofish? (5%)

Answer 8: In a wild-type host (Fig. 3.10.8a) the endogenous *greenwich* (WT) and the two transgenes (T1 & T2) all show rhythmic expression, which reflects the data produced earlier. However, in the *greenwich* null mutant strain (Fig. 3.10.8b) there is no endogenous *greenwich* gene expression and both transgenes show no evidence of rhythmic expression. Therefore the *greenwich* control region requires *greenwich* expression to express in a rhythmic fashion. Therefore *greenwich* expression depends, at least in part, on a feedback loop.

Question 9: Taking each of the five strains in turn, speculate on the type of mutation that could cause the observed *lazy* mRNA and circadian phenotypes. Discuss whether each of the mutations are likely to be dominant, co-dominant or recessive when heterozygous with the wild-type allele. (10%)

Answer 9: Taking each strain in turn:

lazy^{zero1} has no detectable rhythm and no detectable mRNA, so makes no Lazy protein, which presumably explains why it has no rhythm. *lazy*^{zero1} may have a nonsense mutation in an exon that (as long as it's not in the final exon) induces nonsense-mediated decay, or it could have a mutation in the control region that prevents transcription from being initiated. This will probably be recessive with wild-type, although it could show intermediate dominance especially if gene expression is dosage dependent.

lazy^{zero2} has no detectable rhythm but does have rhythmic mRNA at apparently normal levels. Therefore it cannot be a control region mutation that affects transcription. It could therefore be either a missense mutation that results in a protein that has no function, or a nonsense mutation in the final exon, or a mutation that affects translational initiation (such as altering the ribosomal binding site). Again, it will probably recessive with wild-type but could show intermediate or complete dominance if the mutant Lazy protein produced somehow directly or indirectly interacted with the wild-type Lazy protein, or show intermediate dominance if the gene functions in a dosage-dependent manner.

lazy^{zero3} has no detectable rhythm and expresses an intermediate level of mRNA at a constant level. Therefore this is likely to be a mutation in the control region, and the protein, whilst at the wrong levels, is going to be functional. As the level of *lazy* expression seems to be associated with the length of rhythm, this, when heterozygous with wild-type, may be co-dominant with a 'dampened' 24-hour rhythm.

lazy has an extended 30-hour activity rhythm and also has an extended 30-hour mRNA rhythm. This could be associated with lower transcription (meaning a control region mutation), a more stable or less degradable protein (a missense mutation), or a more stable or less degradable mRNA (a mutation possibly in the 3' untranslated region). This could be co-dominant when heterozygous with wild-type, perhaps showing an intermediate 27-hour rhythm phenotype.

lazy^{short} has a shorter 18-hour behavioural rhythm and also has a shortened 18-hour mRNA rhythm. This could be associated with higher transcription (meaning a control region mutation), a less stable or more degradable protein (a missense mutation), or a less stable or more degradable mRNA (a mutation possibly in the 3' untranslated region). This could be co-dominant when heterozygous with wild-type, perhaps showing an intermediate 21-hour rhythm phenotype.

Question 10: The five different *lazy* mutations can generate 10 different transheterozygous combinations of two different *lazy* mutant alleles. Speculate on what their phenotypes might be. (5%)

Answer 10: Taking each combination in turn:

lazy^{zero1}/*lazy*^{zero2}, *lazy*^{zero1}/*lazy*^{zero3} and *lazy*^{zero2}/*lazy*^{zero3} are transheterozygous for two arrhythmic alleles, so would be arrhythmic.

lazy/*lazy*^{zero1}, *lazy*/*lazy*^{zero2} and *lazy*/*lazy*^{zero3} are all transheterozygous for the long rhythm (*lazy*) allele and a null arrhythmic allele. These are all likely to be rhythmic, reflecting the 30-hour rhythm of *lazy*. More specifically, *lazy*^{zero2} makes a non-functional protein in a rhythmic fashion, and this will only affect the *lazy* (30-hour) rhythm if its non-functional protein has some residual function that interferes directly with the Lazy protein of the *lazy* mutant allele, or anything that it binds to. In addition *lazy*^{zero3} seems to make a functional protein in a non-rhythmic fashion, and this may dampen the 30-hour rhythm.

lazy^{short}/*lazy*^{zero1}, *lazy*^{short}/*lazy*^{zero2} and *lazy*^{short}/*lazy*^{zero3} are all transheterozygous for the short rhythm (*lazy*^{short}) allele and a null arrhythmic allele. These are all likely to be rhythmic, reflecting 18-hour rhythm of *lazy*^{short}. More specifically, *lazy*^{zero2} makes a non-functional protein in a rhythmic fashion, and this will only affect the *lazy*^{short} (18-hour) rhythm if its non-functional protein has some residual function that interferes directly with the Lazy^{short} protein of the *lazy*^{short} mutant allele, or anything that it binds to. In addition, *lazy*^{zero3} seems to make a functional protein in a non-rhythmic fashion, and this may dampen the 18-hour rhythm.

lazy/lazy^short transheterozygotes, where one allele induces a long rhythm and the other a short rhythm, presumably have rhythm and may even appear to be wild-type, though the rhythm itself may be dampened.

Question 11: What do these data tell you about the Greenwich and Lazy proteins during the normal 24-hour cycle? (5%)

Answer 11: Both the Greenwich and Lazy proteins show rhythmic changes in protein abundance. Both proteins appear in the late evening (21 hours) and increase in abundance over the next 9 hours or so before decreasing. This more or less reflects the mRNA levels measured earlier (though the initiation and peak of protein levels maybe slightly delayed). In addition, there is a band shift suggesting that the proteins change size, presumably due to some kind of protein modification (such as phosphorylation or glycosylation).

Question 12: What do these data tell you about the effect of light pulses on the Greenwich and Lazy proteins? (5%)

Answer 12: The light pulse has no apparent effect on the Greenwich protein. However, after the light pulse, the Lazy protein disappears, so Lazy is either light-sensitive itself, or light induces a pathway that results in the degradation of the Lazy protein.

Question 13: In the cell culture experiments designed to investigate protein localisation, why were the *greenwich* and *lazy* genes expressed from a constitutively active promoter, rather than their own promoters? (5%)

Answer 13: The objective here is to discover where the proteins localise, so all you want to do is know where the proteins are in the cell. Therefore, you want to avoid any complexities associated with differential expression of the *greenwich* and *lazy* genes, especially as it would be difficult to determine what the cell culture perceives the time of day to be. Therefore constant expression of the *greenwich* and *lazy* genes via a constitutive promoter is the best way to make a cell that has lots of Greenwich and Lazy protein.

Question 14: Briefly explain how deletion mutants, such as those shown in Fig. 3.10.10, could be made. (5%)

Answer 14: They could be constructed in plasmids in bacterial hosts using restriction enzymes. Note that many of the ends are identical which supports the notion that there is something specific about that DNA sequence that makes it a useful place to create an end for a deletion. Alternatively it's possible to use site-directed mutagenesis in a bacterial host. In fact there are several ways in which this might be achieved.

Question 15: Which sequence(s) are responsible for the cytoplasmic localisation of Greenwich protein? (5%)

Answer 15: The three nuclear-located Greenwich deletion proteins all have overlapping amino acid deletions, which implies that the cytoplasmic localisation signal lies in the deleted region that is common to all three, namely between amino acids 444 and 501.

Question 16: How could you characterise more precisely which amino acids are important for the cytoplasmic localisation of Greenwich protein? How could you approach this experimentally? (5%)

Answer 16: To help identify which amino acids are important for the cytoplasmic localisation of Greenwich proteins, you might also look at homology between the Greenwich protein in psychofish, and homologous Greenwich proteins in other rhythmic species. Functionally important amino acids will be conserved between species, indicating that these are the ones that should be investigated first. You could then create more specific mutations in conserved amino acids in regions 452 and 512. In particular, it should be possible to make missense mutations using site-directed mutagenesis by PCR.

Question 17: Propose a model explaining how the *greenwich* and *lazy* genes regulate 24-hour rhythm in psychofish. How would you test your model? (10%)

Answer 17: In the absence of light (at around 18 hours) both *greenwich* and *lazy* are transcribed and translated to make Greenwich and Lazy proteins that are retained in the cytoplasm. They then heterodimerize and the Greenwich–Lazy protein moves to the nucleus where it seems to inhibit *greenwich* transcription (and presumably *lazy* transcription as well, though you have no direct evidence for this). Hence the two genes/proteins are expressed in a rhythmic fashion. The 24-hour nature of the rhythm is in part determined by temperature (10°C appears ideal) and in part determined by the fact that Lazy protein is light-sensitive.

This could be investigated by looking for Greenwich and Lazy DNA binding sites upstream of the *greenwich* and *lazy* genes. You could also look for DNA-binding motifs in the *greenwich* and *lazy* genes.

List of Figures

Chapter 1.4

Chapter 1.5

Chapter 1.6

Chapter 1.7

Chapter 1.8

Chapter 1.9

Chapter 1.10

Chapter 2.1

Chapter 2.2

Chapter 2.3

Chapter 2.4

Chapter 2.5

Chapter 2.6

Chapter 2.7

Chapter 3.2

Chapter 3.3

Chapter 3.4

Chapter 3.5

Chapter 3.6

Chapter 3.7

Chapter 3.8

Chapter 3.9

Chapter 3.10